THE ELUSIVE TRANSFORMATION

SCIENCE, TECHNOLOGY, AND
THE EVOLUTION OF
INTERNATIONAL POLITICS

Eugene B. Skolnikoff

A Council on Foreign Relations Book

PRINCETON UNIVERSITY PRESS   PRINCETON, NEW JERSEY

Copyright © 1993 by Princeton University Press
Published by Princeton University Press, 41 William Street,
Princeton, New Jersey 08540
In the United Kingdom: Princeton University Press, Chichester, West Sussex
All Rights Reserved

Library of Congress Cataloging-in-Publication Data
Skolnikoff, Eugene B.
The elusive transformation: science, technology, and
the evolution of international politics / Eugene B. Skolnikoff.
p.   cm.
Includes index.
ISBN 0-691-08631-1
1. Science and international affairs. 2. Technology and
international affairs. I. Title
Q172.5.I5S47   1992   92-22141
327—dc20   CIP

A Council on Foreign Relations Book

This book has been composed in Linotron Caledonia

Princeton University Press books are printed
on acid-free paper and meet the guidelines
for permanence and durability of the Committee
on Production Guidelines for Book Longevity
of the Council on Library Resources

Printed in the United States of America

10   9   8   7   6   5   4   3

To David, Matthew, and Jessica

_____

Contents _____

Preface

In 1958 I JOINED the staff of Dr. James R. Killian in the White House, shortly after he had accepted the newly created post of Special Assistant to the President for Science and Technology. Since then, I have spent most of my professional life grappling with the subtle and fascinating relationship between international affairs and science and technology. In various guises—first as a government official with James Killian and his two remarkable successors, George Kistiakowsky and Jerome Wiesner, and then since 1963 as a student, teacher, and scholar at MIT—I have tried to understand that relationship. During that time, I have endeavored to write and teach about it in ways that could be useful in the making of policy and the education of students, and to represent it in debates within relevant academic disciplines. Occasionally I have been able to participate in, or advise on, the formulation of policy on relevant issues in the U.S. and other governments and in international organizations, in part through a role in the resurrected White House Science Office under President Carter's science adviser, Frank Press.

As that experience grew, I became increasingly aware that the apparently obvious ways in which science and technology relate to the international landscape are only superficially obvious. Scholarly efforts—and I include my own earlier work—have fallen far short of an adequate probing of the interaction. And it *is* an interaction. Simply laying out the impact of new technology on the relations among nations, the most common approach to the subject, does not do justice to the complexity of the interaction. Rather, to appreciate its full scope and detail requires an understanding of the forces that influence international affairs, a sound grasp of the nature and functioning of the scientific and technological enterprises and of how their outcomes are determined, and a linking of those disparate subjects to bring out the intricate interplay among and between them.

In this study I have set out to present within two covers my conceptions of the interplay of those elements and the conclusions that should be drawn from them. It has been a stimulating and personally rewarding quest, requiring that I explore many subjects and literatures in much greater depth than I had ever attempted before. Some of the conclusions I reached reinforced earlier views, but many were not as I had anticipated. In general, I found that the evolution of the *details* of international politics due to the interaction with advancing science and technology has been as impressive and astonishing as common rhetoric proclaims. But the more general impact on the underlying concepts and assumptions that govern the relationships among nations has in fact been considerably less marked. There are some important exceptions, but they do not counterbalance the realization that fundamental changes in the international system as a whole have been quite limited. This is so notwithstanding the many dramatic alterations of the substance of international politics that have made science and technology major driving forces of international relations, while at the same time tying their progress closely to national goals.

Whether others will agree with this view and its implications I cannot, of course, predict. But I hope that, whether or not there is agreement with this general conclusion, the material here will lead to a sounder appreciation of the detailed nature of the interaction—an appreciation that will serve both practitioners of policy and students of international affairs. My intention is to provide a basis for those who, in a broad sense, are students of the subject, to understand what will influence scientific and technological advance, what are the determinants of future political change, and how it may be possible to cope with some of the more serious issues related to science and technology that will confront the international scene in the future.

I have, in fact, been concerned that scholars and policymakers in international politics too often neglect the intricacies of the role of technology. Rhetoric and lip service to that role abound, but rarely is there a serious attempt to deal with the dynamic nature of the scientific and technological system—a subject not as impenetrable as is often assumed—and with the breadth of its relationship to international politics. If this book helps to overcome that neglect, it will have met one of its central purposes.

The breadth of the material and literatures that must be considered has often deterred attempts at a comprehensive approach to this subject. Science and technology permeate, to varying degrees, almost every significant issue in international affairs. It is not possible to do justice to them all, nor is it necessary. But all significant aspects do have to be encompassed if the interaction is to be seen in its full, robust character. Perhaps unwisely, I have attempted to do just that, distilling from each issue what I believed to be the salient elements and generalizations.

The result is a study that relies heavily on literatures in a wide variety of fields and that necessarily also draws on current events (up to the spring of 1992), as well as on my own experience and research. In the end, it is very much a personal statement. Even the material drawn from evidence and analysis of other scholars and from policymakers reflects my judgments about their significance and about the lessons to be learned. Whatever measure of agreement this book finds, the search for deeper understanding of the relationship was for me a fascinating learning experience. My predominant hope is that it will inspire or provoke interest and response from others.

There are many to acknowledge for the help they have given in the process of carrying out and writing this study. Most have helped unwittingly in the course of the many years in which I have worked in this particular vineyard; they include colleagues at MIT and in Washington, fellow scholars at meetings and conferences, and the many from outside the United States with whom I have worked or debated. But a few have contributed in special ways.

The most striking is one individual, Harvey Brooks, whose influence in almost all aspects of the interaction of science and technology with public affairs has been simply unparalleled. Now an emeritus professor at Harvard, Harvey Brooks has published more widely, and more wisely, on these topics than anyone else; and he is unfailingly generous with his time to review the works of others. In my reading for the many different issues discussed in this study, I had the experience, common to many, of finding that the best summary piece, and often the one that made just the point I was searching for, was an article written by Harvey Brooks. Hence, there are frequent citations of his work throughout the pages that follow.

But that does not encompass my debt to Harvey Brooks. We have worked together in a variety of settings and on several articles over many years, and we have taught a joint Harvard/MIT course together. On all of those occasions, I have benefited directly from his knowledge and wisdom. On this study, I have had the advantage of his careful and invaluable review of an earlier draft of this study. He cannot be tasked with the conclusions I reach or with any errors that have crept in, but at the same time he cannot escape the fact that his influence has been pervasive for many years.

There are many others who contributed more than casually to the often-tortuous progress of this study, which extended over more than five years. Joseph Nye read a very early draft while we were both on sabbatical in Oxford, and he provided an important set of comments and timely support for the ideas that were then emerging. Kenneth Keller at the Council on Foreign Relations took a lively interest in the work and strongly encouraged the Council's attention to the subject and to the book as it was developing. Lewis Branscomb, along with Harvey Brooks, cochaired several substantive and lively Council sessions focused on the book at the beginning of the project in 1988 and then again as it neared completion in 1991. These sessions were important in the development of many of the ideas, and I offer a collective thanks to the participants.

Several colleagues at MIT, Oxford, and elsewhere read various drafts and provided helpful comments—in particular Nazli Choucri, Richard Samuels, David Victor, Ted Greenwood, Mitchel Wallerstein, Peter Katzenstein, the late John Vincent, and students in classes, who had to endure the exploration of unfinished ideas but who provided valuable reactions and debate. Amelia Leiss, Associate Director of the MIT Center for International Studies, was an invaluable support as this work got under way, continuing our long and productive professional relationship. Guido Goldman, a warm friend of many years, provided a beautiful setting in Maine for the final hectic steps in completing the manuscript. Lawrence Aronovich and Caren Addis made important contributions as research assistants at various stages of the project, as only excellent graduate students can do, and Carol Conaway stepped in with valuable bibliographic help near the end. And, even in the computer age when manuscripts are typed entirely by authors, I want to acknowledge the assistance of my secretary, Parth Domke, who helped greatly in making it feasible to complete the last stages of the manuscript while I was carrying a full academic load.

I was also extremely fortunate to receive support throughout the project from several foundations that made the extended work feasible, and even enjoyable. The Sloan Foundation provided the initial grant, offering not only necessary funds, but also a vote of confidence in me and in the subject. In particular, Arthur Singer of the Foundation staff is to be thanked for his initial support and Michael Teitelbaum for his continuing responsibility for the grant. I know I speak for many authors for whom Arthur Singer is an unsung hero because of his and the Foundation's help at a crucial time in their careers. I also received assistance at various times from the Carnegie Corporation, from the Japanese Endowment for Energy Policy Studies at MIT, and indirectly from the MacArthur Foundation through its support of the Program on International Peace and Security at the MIT Center for International Studies. And, in general, the stimulating and supportive environment of the Center and of MIT as a whole was of great help throughout the project.

And I will not forgo a note of deep appreciation to Winifred Skolnikoff, who put up with the often-surprising demands on the author's time and attention that a project of this kind seems to require. I am sure that she did not enjoy the enterprise as much as I did, but her help in innumerable ways was indispensible to its completion.

I also benefited from two other opportunities that were essential to the project. The first was a semester sabbatical at Balliol College, Oxford, under the MIT/Balliol Faculty Exchange program established by William Coolidge, a graduate of both institutions. That was a superb experience that allowed me, in the unparalleled environment offered by Balliol, to write the first draft of most of the study. The other was the opportunity provided by the Rockefeller Foundation to spend a month as a scholar-in-residence at the Foundation's Bellagio Study and Conference Center. That inspiring setting, the fellowship and stimulation of other scholars from many different fields, and the hospitality offered by the staff provided the means to consolidate the ideas of the body of the work and to formulate the concluding chapters. It would not be possible to exaggerate the importance of those two experiences in the writing of this book.

With that kind of support, and the opportunities it afforded, any faults in the work that follows can only be due to the author's fallibility.

PART ONE

PART ONE

One

The Setting

THE STARTLING CHANGES in world affairs that began in the late 1980s signaled the end of many of the central elements of postwar international relationships. Momentous and unexpected events in Eastern Europe and the Soviet Union occurred at a breathtaking pace, with a spontaneity that tended to obscure the underlying currents that had unleashed them. Many forces were at work in those societies over decades, culminating in dramatic upheaval in essentially all countries of the former eastern bloc.

Not the least of those forces were the effects of technological change on the nations of the East and on their relations with nations outside the bloc—nations that were themselves heavily influenced by the pace and nature of advances in science and technology. The influences of technological change were both subtle and dramatic, ranging from the effects of the irresistible flood of information crossing borders to the domestic structural changes required in order to be competitive in high technology; from the change in the cost and rationality of the use of military force to the consequences of the rapid growth of the service sector in national economies; from the recognition that a nation could not conceal the effects of a nuclear accident from the outside world to a realization of the great difficulty of coping with environmental degradation in a centrally planned economy.

The influence of technological change in the disintegration of communism in Eastern Europe and the Soviet Union is but one manifestation of a much larger story in which the results of science and technology have contributed to a profound evolution of the details and substance of national and international affairs. The effects are visible not only in the outcome of the communist experiment, but in the countless alterations in the relationships within and among nations and peoples. And that influence is likely to continue to be significant long into the future as the nations of the world remain strongly committed to supporting research and its products.

The evidence for the role of science and technology in the evolution of world affairs is pervasive, and most easily seen through dramatic developments that have had global consequences—such as the deployment of massive strategic nuclear forces, the nuclear accident at Chernobyl, foreign-currency transactions on computerized financial markets in excess of $500 billion per day, or the total eradication of the scourge of smallpox. Other developments are less immediately spectacular but as far-reaching:

the ability to fax documents and to reach individual telephones instantaneously and inexpensively throughout the world, the increasing dependence of modern weapons systems on technology and on technological advance, the relevance of technological competence to a nation's competitive position, or the immediacy and global reach of television. Still other developments serve to underline the more intensive relation between local actions and global consequences: degradation of the stratospheric ozone layer due to the widespread use of chlorofluorocarbons, the consequences of disruption in energy supplies or failure of information systems, or the climatic effects of the accumulation in the atmosphere of the effluents of energy technologies.

The relationship between international affairs and science and technology is not only a result of recent advances in science and technology, though the breadth of interaction and the rapidity of change are relatively modern characteristics. The historical examples are countless, ranging from weapons developments that altered the fate of nations and social structures, to industrial technologies that were the basis of revolutions in economy and wealth, to new capabilities in science and technology themselves that led to astonishing discoveries and applications.[1] It was not only the physical developments of technology that had an impact; underlying concepts of science and of the natural world were crucial as well. The intellectual currents of the Enlightenment, so much a product of the new ideas of experiment and rationality that accompanied the scientific revolution, served to stimulate massive forces for change in the West. The French and American revolutions were products of those forces, as were the concepts embodied in the American constitution.[2]

Today, given the scale and organization of the scientific and technological enterprises, science and technology have become arguably the most powerful and persistent factors leading to societal change and, necessarily, to change in international affairs. The accelerated commitment of resources to research and development (r/d) during and after World War II has transformed the relatively haphazard climate of invention and scientific research of earlier centuries. There is now in place a formidable and growing capacity—a system—for targeting human ingenuity toward the rapid expansion of knowledge and the production of new technologies designed to serve perceived or speculative needs. Not only do the products of this system have significant international effects, but its very operation leads to international consequences and favors the creation of global markets for its products; and, for a surprising portion of the system, international goals provide the underlying motivation for the commitment of r/d resources by governments and even by industry.

The subjects with international consequences that have been massively affected by technological change in recent years are familiar: weapons, communications, the economy, transportation, agriculture, health, space, and others; few, if any, aspects of international affairs have been untouched by science and technology. The effects of the application of technology are so widespread, in fact, and often of such obvious importance in the conduct and evolution of relations among nations, that it is routine for commentators to lapse into florid rhetoric in describing the resulting dramatic change in the international political system. Secretary of State George P. Shultz said in December 1987, "Developments in science and social organization are altering the world profoundly—too profoundly for conventional habits of thinking to grasp. History suggests that mankind rarely understands revolutionary change at the time it is coming about." W. Michael Blumenthal, Secretary of the Treasury under President Carter, said in his 1987 Elihu Root lectures at the Council on Foreign Relations in New York, "I believe there is one circumstance which overshadows all else and has set the current period apart: unprecedented, deep and continual technological change . . . extraordinarily rapid technological change has thrust upon us new and as yet unresolved problems of governance in the national and international spheres."[3]

For all that the relationship appears to be self-evident, the extent even of the surface changes in international politics stemming from science and technology proves to be quite difficult to characterize with precision or to assess satisfactorily. It is not hard to draw up lists enumerating international political issues that are affected by technological change, but it is very much more demanding to understand the complexity of the interactions and their more profound consequences for international affairs. That understanding is necessary not only to capture the relationship for analytical purposes, but also to assess the true extent of the evolution in international affairs—and to be able to anticipate, and possibly influence, the future directions and implications of change.

It is that understanding of the complexity of the interaction and of the consequences for international affairs that we hope to achieve in this inquiry.

Assessing the Interaction of Science and Technology with International Affairs

The most common approach to studying the interaction of science and technology with international affairs views the relationship in the context of specific policy areas, typically in relation to pressing policy concerns. A limited number of studies have taken a somewhat broader canvas using a variety of specific policy implications of science and technology as a way of illustrating the growth and change in the subject matter of international politics and the new relationships and institutions that have been cre-ated.[4] These studies have been useful with respect to specific policy impli-cations of developments in science and technology, but they are less satis-factory for our goal of understanding the broader and more fundamental interactions of science and technology with the international political sys-tem, and how that system is affected by the continued advance of science and technology.

Our primary purpose is not to produce a definitive or quantitative mea-sure of system change; the task would be difficult and the result ultimately arbitrary. Rather, we will explore the nature of the interactions between technological and social factors that lead to evolutionary change, identify the direction and patterns of that change, and record its characteris-tics.[5] Our focus, accordingly, will be on the patterns of evolution of impor-tant elements of international affairs as a result of the interaction with science and technology; we will consider change in system characteristics or concepts to constitute a definitive transformation only when it is un-ambiguous.

Within that constraint of identifying patterns of change rather than ab-solutes, we will be asking how traditional geopolitical factors, such as mil-itary force, natural-resource endowment, and economic capacity evolve in meaning as they are affected by, and interact with, technological devel-opments. We will want to know if, and in what ways, new or altered fac-tors—such as scientific and technological competence, technological com-petitiveness, effectiveness of international organizations, decision-making capacity in a technological world, and the greatly expanded international flow of information—have now become significant in international rela-tionships. And we will want to understand how more basic elements, such as governmental authority, relations of dependency, and military and eco-nomic power, have been altered in significance and content by the effects of science and technology. Central to the analysis will be the exploration of how the characteristic outcomes of the scientific and technological sys-tem affect the evolution in these and other elements of international affairs, and are in turn influenced by them.

The larger question of whether change induced by science and technol-ogy constitutes revolutionary rather than just evolutionary change in the international system cannot help being of great interest, notwithstanding the difficulty of measuring change. James N. Rosenau, in a recent and provocative book, argues that a fundamental transformation in world poli-tics *has* occurred, with the dynamics of technology serving as a major driving force. His argument is that breakpoints or discontinuities occur when the primary parameters of the international system have been trans-formed; and he believes and seeks to demonstrate analytically that such transformations have now taken place, largely as a result of the effects of technological change.[6]

There is no doubt that international politics is quite different, in almost all dimensions, than it has been, or than it will be. It is evolving under the influence of technological advance perhaps faster than ever before. The more telling observation, however, is the persistence and adaptability of traditional concepts in the face of rapid evolution. Technology-related changes may be modifying the dimensions of national autonomy but not the assumptions of autonomy in national policies, changing the substance of dependency relationships but not the fact of dependency, altering the nature of weapons but not denying a role for power in international affairs, modifying the distribution of power and capabilities but not the significance of those attributes of states, creating new patterns of economic interaction among societies but leaving the management of the economic system largely in national hands, altering the relationships between government and nongovernment actors but not the basic authority of governments, raising wholly new issues and altering traditional issues that must be dealt with internationally—but thereby making foreign policy more complex, not fundamentally different.

Rosenau would argue that the persistence of traditional concepts does not reflect the reality of the changes that have taken place; that, in fact, the rise of what he calls a "multi-centric" world parallel to and overlapping the state-centric world has overwhelmingly reduced the predominance of states in the actual conduct of international relationships.[7] Certainly there has been a relative diminution of state power over the details of international activities, a fact we will examine in part 2 of this book; but whether it is as extensive as Rosenau concludes, or as decisive politically, is a more debatable proposition. We will not try in this study to prove whether or not there has been a system transformation, for that ultimately can only be a matter of definition or *post hoc* evaluation. We will, however, attempt to lay out clearly the dimensions and nature of the evolution that results from technological change so others may form their own judgments.

My own judgment on this larger question of system transformation is that the fundamentals of the nation-state system have not been altered as much as most rhetoric would lead us to believe; and that they will not be materially altered in the foreseeable future, though much on the international scene will change. As we will see, there are some specific and important changes in these fundamentals, but not enough, in my view, to invalidate the overall conclusion. By fundamentals of the system I mean simply the congeries of independent states, each jealous of its independence and seeing itself in competition with others, each attempting to maintain freedom of action and each committed to maximizing national welfare and influence.

If my judgment that system transformation has not taken place is correct, it is appropriate to ask whether in the future the international system—having approximately the same structure and assumptions as at present—will be able to meet the challenges that will be posed by the effects of continuing technological change, especially those mandating extensive and politically intimidating collective action. Will assumptions built into the nation-state system about goals, prerogatives, freedom of action, and independence be compatible with those challenges? Can nations acting in their own interests be expected to cooperate to the degree that may be required in the face of high uncertainty and high cost, even if the necessity and benefits of cooperation are evident? What would be the effect on the international system if the present structure proves to be inadequate?

The analysis in this study will lay the groundwork for judging both the likely scale of future demands on the system related to technological change—many of which will arise in an ecological context—and the probability that a systems transformation will be necessary or probable to meet those demands. Though prognostication is not my primary intent, my conclusion is that neither through necessity or intent, except possibly in the face of impending (and credible) global catastrophe, could there be a substantial systems transformation in the foreseeable future. In fact, it is more likely that the fundamentals of the present system of nation-states will prove to be more, rather than less, necessary to manage the increasingly difficult future international environment that technology will help create.

We will return to these questions in the concluding chapter, when the evidence is before us.

Primary Questions

Thus, our primary goals are to understand both the comprehensive interaction of science and technology with international affairs, and the central consequences of that interaction for international affairs. We will do that by considering two broad questions.

1. What characteristics of the modern scientific and technological system are important in the evolution of international affairs, and can their effects be generalized?

2. How have significant elements in international affairs evolved as they interact with technological change, and in particular have underlying assumptions and concepts been altered?

As we develop evidence on those questions, we will want to keep in mind the judgmental question of whether a larger transformation in international politics due to technological change may be said to have taken place, or be in train. If not, can it be said that the evolution already experienced and the expected effects of technological change on the horizon will eventually lead to a more fundamental reordering of the nation-state structure of the international system?

Some Comments on Theoretical Issues and Literature

Curiously, though there have been innumerable policy studies dealing with the effects of technological change in specific policy areas, and a small number of works that attempt to look across the board at important policy areas, the subject is largely unexplored from the deeper perspective of the overall effects of science and technology on the evolution of international affairs. It is curious because of the evident centrality of the relationship, whether or not there is agreement on its ultimate significance. Even scholars concerned with theoretical issues in international relations tend to treat science and technology as static "givens," or as emanating from impenetrable black boxes.[8] One of the purposes of this study will be not only to fill this lacuna in the subject as a whole, but along the way to provide a basis for considering science and technology more appropriately in policy or theoretical analyses—as the interactive, dynamic variables they are in reality.

Notwithstanding the relative paucity of theoretical work on the overall subject of this study, there is obviously much in the theoretical literature on international relations that is pertinent and that would illuminate the issues we will be discussing. It would be possible, in fact, to structure this book along the lines of the debates among theorists—for example, according to differing views of the role of the state or of international organizations, both of which have been, and will continue to be, much affected by scientific and technological change.

I have elected to proceed along a different path, for this study is not intended to be only or primarily a contribution to theory; a structure based on current theoretical concepts would tend to obscure both the critical sense of the dynamic nature of technology and the generalizations that cut across many different policy domains. We will return, however, in chapter 7 to explore some of the implications of the analysis for international-relations theory. There will, as well, be frequent reference to, and testing of, theoretical ideas in the body of the work as appropriate.

The idea of sovereignty, appearing often in both theoretical and policy terms, does require brief discussion, however, for it is a central element in the nation-state system and is repeatedly cited as having been eroded or at least greatly altered in meaning by technological change.[9] It might easily be assumed that it would figure as an organizing theme in a study concerned with the effects of science and technology on evolution in international affairs.

The concept is generally thought of by theorists, statesmen, and the public at large as a fundamental attribute of nations, what Stephen Krasner calls "the constitutive principle of the existing international system."[10] However, sovereignty has clearly become either a term of art for theorists or a commonly used but imprecise idea that carries heavy emotional baggage. Definitions of sovereignty, even as a theoretical concept, vary substantially. Some treat it as an absolute value, in effect coincident with the concept of an independent state; either sovereignty exists or it doesn't, and there is no middle ground.[11] Others have a much more flexible definition, encompassing notions more in accord with commonly understood ideas about the extent of state autonomy and control.

Rather than attempt to structure our analysis around theoretical concepts that, even with careful definition, tend to mean different things to different readers (that problem of definition is already severe even for the terms *science* and *technology*, as we will see), it will be wiser to leave the discussion of effects on sovereignty to the final chapter, when the evidence of the analysis will make greater precision possible. Accordingly, in the body of the study we will avoid the use of the term *sovereignty*, preferring instead to speak of the freedom of action of governments or of changes in the authority of the state.

Science and Technology as Causes of Societal Change

The manner in which developments and applications of science and technology come about and interact with other social factors to lead to societal change clearly must be a part of our inquiry. That subject will be taken up in chapter 2, but some preliminary remarks are needed to refine the use of words dealing with causation, effects, and change, both social and technological. This is necessary primarily to avoid repeated use of cumbersome qualifying language, but it also is important to clarify concepts not well defined in everyday language.

First is the perception of "causation." Science and technology, or more accurately scientific developments and the applications of technology, do not "cause" changes in the international political system or in any societal elements. The effects that emerge as those developments and applications are embodied in social structures may alter the relationships among structural factors, make new actors relevant, modify the framework of policy issues, alter perceived costs and benefits of policy intervention, and lead to other modifications in structure and context. But these changes come about through the actions of the actors in the political and economic system, through the choices they make, through the cumulative responses of society; they do not come about simply because of the existence of the new scientific knowledge or technological capabilities.

Nor are the effects of the introduction of a technology necessarily the same in different societies; the intermediation of different social systems may, and usually does, lead to quite different patterns of assimilation of technological change. Science and technology are factors that must be included in the analysis of societal change; their developments and applications may even prove to be forcing factors that lead to a societal response. But they are still only elements of an issue, to be considered along with other political, economic, and social factors.

This issue of causality is closely tied to the question of technological determinism—whether the advance of technology is logically inevitable and effectively independent of conscious choice. The view that technology is by nature deterministic leads to the conviction that the outcomes of the scientific and technological enterprises will have to be dealt with as though human choice had little to do with producing those outcomes, as though the arrow of causality points in one direction only: from technology to societal effect. How technological outcomes are determined is an important part of our inquiry that will be treated in detail in the next chapter.

But it is worth noting here that even though technology is in fact a product of human choice, in practice it has to be treated both as a dependent and as an independent variable. Technological change comes about as a result of human decision and is in that sense a dependent variable. But society is also often confronted with new situations in which technological change brought into being for one purpose has consequences in other and broader areas—for which it is, in effect, an independent variable. This is one of the results of the diffuse nature of decision making in the development and application of technology, and the often unplanned or unpredicted side effects resulting from its interaction with other social factors. Thus, there is a sense in which the frequently expressed idea of the "momentum of technological development," implying a force not in human control, is an appropriate characterization, even if not strictly accurate.

In common discourse it is usual, even standard, to talk about technology as "bringing about" or "causing" some international change, whether it be new degrees of interdependence or the emergence of superpowers. That phraseology is not accurate unless adequately qualified; but it is awkward, at best, constantly to have to provide the qualification, and often unnecessary to the point being made. It is simply sophistry to quarrel with an assertion such as "technology has changed the meaning of major war," unless it is in the context of a detailed discussion of, for example, social choice in the development of weapons technology. We will try to avoid imprecise statements of causality, but, with apologies to purists, the shorthand idea of technological causation will occasionally be used simply to avoid repetitive qualifications. It will never, of course, be used when a more precise statement is necessary for the argument.

A further complication of language follows from the fact that it is not developments of new knowledge in science and technology that are the relevant changes, but rather their applications or embodiments in a socio-economic system. Again, common usage usually refers only to developments in science or technology and tends not to make that distinction clear. The more extended and accurate phraseology will be used here much of the time, but the shorthand reference to science and technology will occasionally be allowed for the sake of simplicity.

It is also sometimes assumed, rhetorically at least, that societal change takes place quite suddenly as the result of the introduction of a new capability, often technological, such as a new method for producing energy or a new weapons system. In fact, societal change does not occur in sudden discontinuities in response to such developments, but gradually over time. Though this is partly a matter of the definitions of "sudden" and "gradually," the important recognition is that social effects normally evolve through the cumulative impact of incremental developments, in which technological change is but one factor. Large-scale war or a natural catastrophe can result in rapid alteration of social relationships and structures, but new technologies—even seemingly radical ones such as nuclear weapons—lead to social effects through an evolutionary process of social learning and adaptation, rather than through sudden, quantum shifts of attitude, behavior, or structure. The technologies themselves only diffuse slowly through society, with social impact occurring only after their wide dissemination (nuclear weapons may be something of an exception, though even that is arguable). In fact, initial assumptions about how quickly society will respond to dramatic technological developments often prove over time to be quite different from what actually happens. Societal change today may be more rapid compared to the past, but it is rarely discontinuous.

What Is Meant by "Science" and "Technology"

We are interested in both science and technology, but science is not always mentioned in the paragraphs above. That is not an oversight or a device for economy of expression. Technology is, in fact, the more prevalent of the two in their interactions with international affairs. Science, and scientific research, is less of a direct factor, though there are important exceptions. Science is, of course, important in technological development, and it is also central to the ability to assess the social and other implications of the application of technology. The realization that there are major problems in, for example, the earth's ozone layer or in the buildup of carbon dioxide in the atmosphere is a direct product of scientific research (itself aided and abetted by the technologies of computers, instruments, and space vehicles, among many others). And science and scientists have a direct role in transnational relations.

Nonetheless, the more extensive relationship is between technology and international affairs, so that our primary focus will be on technology. Science is *not* meant to be subsumed under the word technology, except insofar as it is an intrinsic part of technology itself. When science or particular aspects of science are directly relevant to the discussion, they will be explicitly included.

The immediate need is to define what is to be meant by "science" and "technology," a task more daunting, particularly for the latter, than might at first be thought. The problem is that the word technology has evolved in its implications, probably roughly in parallel with the growth of the significance of technology in human affairs. Moreover, it has become a focus of analysis in a variety of philosophical and social literatures, in which it acquires elaborate and precise definitions. At the same time, the lay public—and some writers who should know better—often broaden the meaning of "technology" to include science within its compass (or the reverse).

A highly precise definition of "technology" is not essential for the purpose of this work, but a good, working boundary for the concept is.[12] One of the prime sources of confusion, aside from the issue of the relation of technology to science, is whether technology should be thought of as a piece of physical hardware, which tends to be a layman's offhand view of the meaning, or whether it should rather refer to the knowledge base that made the hardware possible. Is it physical hardware only, or also the knowledge, techniques, and procedures for carrying out tasks, which would include, for example, management? Could it be disembodied knowledge, or must it be seen in relation to the accomplishment of specific purposes in order to distinguish it from science?

I find the most useful conceptual definition for this study to be that given by Harvey Brooks, who has defined technology, in a piece we will return to somewhat later for other of its insights, as "knowledge of how to fulfill certain human purposes in a specifiable and reproducible way."[13] The conditions of reproducibility and dedication to fulfilling human purposes are what separates technology from art and from science.[14] Elaborating on this definition, Brooks notes, "Technology . . . does not consist of artifacts but of the . . . knowledge that underlies the artifacts and the way they can be used in society."[15] This may at first seem counterintuitive to some, but it is critical to understand that technology, at least as the term is used here, implies more than simply a piece of hardware; it implies the larger knowledge base that is specific to the creation of a particular piece of hardware and that made its production and application possible.

It is also important to recognize that the focus on specifiable knowledge, not artifacts, indicates that technology legitimately includes bodies of rules and techniques for accomplishing purposes, just as it includes hardware. What might be termed "social" technologies, such as codified systems of management or computer software, are therefore appropriately considered technologies along with those that are physical in nature.

Innovation, on the other hand, is the process by which technology is created and deployed in society, implying the creation as well of whatever support systems are necessary to install and use a technology. Edison's inventions were of little social significance on their own; they became technologies as the knowledge of how they operated became reproducible, and they became innovations when support systems—including other inventions—became available: electric-power grids, home wiring, accounting and sales bureaus in commercial organizations, and other elements of systems able to deliver power to customers.[16]

Science can also be defined in a great variety of ways, though perhaps the variance is not as large as it is for technology. The most useful approach for now is to emphasize the contrast with the definition of technology. In this formulation, science would be the knowledge of "how and why things are as they are," as opposed to knowledge of "how to fulfill certain human purposes in a specifiable and reproducible way."[17] Thus, it refers to the accumulated body of knowledge of the natural and social world, without necessary reference to the application of that knowledge for human purposes.

The words science and technology both have an inherent ambiguity of usage, sometimes referring to knowledge itself and sometimes referring to the organized process for acquiring knowledge. The word technology is perhaps not so frequently used to mean a form of organized human activity—in contrast with "engineering," which always has that implication. "Science," on the other hand, is intended to mean the process for the conduct of research at least as often as it is the results of research. In the course of this study, the intended meaning will be made clear whenever the reference might otherwise be in doubt; the idea of the scientific or technological "enterprises," or systems, will often be used when the focus is to be clearly on the process as opposed to the results.

In practice, the boundary between technology and science is often not as clear as those definitions would imply. What is clear is that technology is not necessarily based on prior progress in science; in fact, until recently it was more likely to be based on prior technology. The relation between science and technology is complex, contains many feedback paths, and cannot be characterized by a simple linear progression from science to technology to application, even though a higher proportion of technology is science-based today than in earlier eras. The relationship between them, as we will see, also has significance for international affairs.

The social sciences and social technologies are necessarily relevant to this inquiry, for they fall within the definitions we have just presented of science and technology. They are relevant because they are concerned with, interact with, and have an influence on international affairs. And, most persuasively, they are relevant because the approach and much of the data of this study are deeply embedded in the social sciences.

Nevertheless, to avoid excessive diffusion of the analysis, social sciences and social technologies will not be major subjects of the study; they will not be excluded, but neither will they be given equal treatment. Though there are many similarities of substance, and even overlap, between the social and physical versions of science and technology (the process of innovation by definition involves social factors along with physical ones), focusing on the social versions would introduce the need for additional historical accounts, discussion of many ambiguities, and presentation of controversies that would complicate an already-complex story. The primary interest of this study is the relation of the natural sciences and technology to international affairs, notwithstanding the importance of both the natural and social sciences.[18]

Plan of the Study

We will present first, in chapter 2, a picture of the scientific and technological enterprises as they have evolved to the present day, with an analysis of the driving forces behind them, the nature of their internal structure and complexity, and the character of their outcomes as conditioned by internal structure and external pressures. The following three chapters will explore the evolution of international affairs under the influence of those outcomes, with the universe of issues divided into three analytical categories: national security, economies and polities, and global dangers. Chapter 6 will identify the implications for the processes of national and international governance, so critical an aspect of evolution of the international system. Finally, chapter 7 will bring together the general conclusions, point up the relevance to theoretical debates in the literature, and draw the larger systemic lessons that emerge.

Two

The Scientific and Technological Enterprises and the Direction of Technological Change

Historical Evolution

The scientific age can be said to have begun in the seventeenth century with Francis Bacon's recognition of the significance of a disciplined method for development, testing, and verification of theory. His concept became the essential foundation for all that followed by recognizing the cumulative nature of science that transcended the capability of individuals or of a generation.[1] The British Royal Society propagated the Baconian doctrine, providing the base for the development of a scientific community that steadily expanded as succeeding generations built on what went before. It was that adoption of the scientific method that distinguished science in the West from science in the Islamic world or China, both of which could be considered to have had scientific and technological accomplishments in the fifteenth and sixteenth centuries well in advance of those of Europe.

The achievements of the scientific community in the West over succeeding centuries amply demonstrated the vitality of this approach to the study of natural phenomena. Greatly aided by technological developments such as the printing press, the telescope, the microscope, accurate clocks, and countless others, science and scientific research became a major intellectual activity throughout Europe, spreading in time to the New World and to the nations of the East. Those technological developments so essential for science and for the economic growth of the West, however, were not themselves the products of research. Some preceded the origins of the new science, others were developed without scientific understanding of how or why they worked; many, in fact, provided the impetus for development of new scientific fields. It took almost three hundred years before science became an important factor in commerce or war—that is, before the results of the laboratory became the basis for substantial practical innovations.[2]

Instead, technology advanced in an intimate relationship with economic growth. The historical question of which was cause and which was effect—economic growth or technological innovation—need not concern us; the important observation is that the two have gone hand in hand since the late Middle Ages.[3] There may be good reason to believe, however, that economic growth today is more clearly dependent on successful technological innovation than the reverse, a point to which we will return.

The environment that made economic and technological growth possible had many and varied components. Nathan Rosenberg and L. E. Birdzell, Jr., single out the growth of three closely intertwined elements as critical: autonomy, diversity, and experiment. Autonomy was a result of the weakening of political and religious controls, a weakening that allowed individuals and organizations to challenge existing patterns of power and structure, to experiment, and to take risks. Diversity grew out of the effects of autonomy, as institutions able to use a variety of resources and to cater to differentiated needs came into existence. Experiment was an essential aspect of the other two, for it admitted the possibility of change— change in organizations and change in technology; and it implied both the rewards of success and the risks of failure.[4]

Paul Kennedy draws a roughly similar, if less precise, analysis in his explanation of the "European miracle," putting more weight on the evolution of economies and institutions but omitting the crucial role of experimentation. In particular, he singles out the competition that resulted from the fragmentation and decentralization of power and decision making in Europe, the development of sustained and predictable market and other institutions, and the availability to all nations of the rapidly changing technology of armaments. He argues that "there was a *dynamic* involved, driven chiefly by economic and technological advances, although always interacting with other variables such as social structure, geography, and the occasional accident."[5]

The results led to the exponential growth of the economies of the West as a whole, with technological innovation a central element. But not all nations of the West proved to be equally congenial at any given time to what was necessary for technological and economic growth, so that the influence of factors relevant to a nation's international status ebbed and flowed at different rates from one era to the next. It is worth quoting Paul Kennedy's key paragraph:

> The argument in this book has been that there exists a dynamic for change, driven chiefly by economic and technological developments, which then impact upon social structures, political systems, military power, and the position of individual states and empires. The speed of this global economic change has not been a uniform one, simply because the pace of technological innovation and economic growth is itself irregular, conditioned by the circumstance of the individual inventor and entrepreneur as well as by climate, disease, wars, geography, the social framework, and so on. In the same way, different regions and

societies across the globe have experienced a faster *or* slower rate of growth, depending not only upon the shifting patterns of technology, production, and trade, but also upon their receptivity to the new modes of increasing output and wealth.[6]

This relationship between the technological dynamic of change and the "position of individual states and empires" is at the heart of our study. What was valid in the past is still valid, and in fact more significant as the breadth and rate of technological change continue to increase.

By the end of the nineteenth century, the role of science had become more closely tied to application, whether for wealth or for war. Industrial research laboratories began to appear in the last quarter of the century, particularly in the chemical industry. They were established as it became necessary (or advantageous) to commercial exploitation to understand why things worked as they did, and in part as the subject matter of technology—electricity, magnetism, chemical processes—came to require specialized knowledge and experiment not accessible even to the experienced artisan without benefit of more systematic technical training.

Gradually, and then more rapidly in the twentieth century, this introduction of structured research and experimentation in industry in the interest of innovation became a self-sustaining system, one unique to the West at the time. To judge by its results, it was one of the more important institutional developments of this century, serving to accelerate the rate of change of technology and thus of change in society. In the words of Rosenberg and Birdzell: "It improved recognition of the possibilities of change, reduced the risks of attempting change, and increased the probable rewards of change. It thus altered the goals and incentives of Western economic systems toward more change and growth."[7]

The other key institutional development of the twentieth century that provided an enormous stimulus to technological innovation was the growth of the role of governments in support of both science and technology, at first primarily for security purposes. This was not a new role, for states had long been interested in technological innovation for application to armaments and the waging of war; but they had been interested predominantly as potential customers of independent developments rather than as paymasters for development of a technology. Even then, governments were not always willing consumers when new technologies threatened existing military organizations or tactics. Steam power was long delayed in being applied to navies in the nineteenth century because of resistance from senior officers—as the adoption of the bomber was delayed in this century by the U.S. Army, and even the ballistic missile by the bomber-equipped air force in more recent years.[8]

But in this century—at first modestly and then with savage intensity during World War II and the years that followed—governments have devoted huge sums to the support of science and technology for military applications. The interwar years saw a substantial advance in the knowledge that could be applied to weapons in fields such as aircraft, radar, mines, and tanks and to the materials and electronics that backed them up. When the war came, the commitment made to the scientific and technological enterprises for war purposes on both sides had no earlier parallel. The developments in nuclear energy, radar, proximity fuses, V-2 rockets, bombers, bombs and bombsights, communications, intelligence, materials, organization, operations research, and countless other fields were all supported largely with government funding. The work was usually carried out in new or wholly transformed organizational forms and units and involved thousands of scientists and engineers who had no previous knowledge of the requirements of warfare. The enterprise was entirely unique, and quite spectacular.[9] It is a fair judgment that this was the first war in history in which the scientific and technological developments achieved *during* the war had an effect on the outcome of the war. They may even have been the major factor in determining that outcome.

It was not only in the furtherance of military goals that governments moved increasingly toward direct support of science and technology. There had been a long tradition of the commitment of funds from public treasuries for scientific activities that would redound to the greater glory of the nation or the monarch, or that might have pecuniary or other useful results. The Age of Exploration saw many expeditions launched to advance knowledge, as well as to claim territory and discover new sources of wealth. Individual scientists in Britain, France, and Germany in the seventeenth, eighteenth, and nineteenth centuries often were provided with funds by patrons, some of whom were monarchs, government officials, or state ministries, frequently acting with quite mixed motives. At the beginning of the nineteenth century, President Thomas Jefferson engaged the young United States of America in the support of the Lewis and Clark expedition to explore, record flora and fauna, and map the northwest of the continent. The purpose was seen as an investment for the future commerce of the nation, and not incidentally to establish a territorial claim; science was in effect the instrument to accomplish those practical and political aims.[10] Almost a century later, the United States embarked on a seminal commitment to the support of agricultural research (and transfer of this knowledge for use on the farm) that provided the knowledge and experimental base to make agriculture in the United States the most productive in the world for the next one hundred years.[11]

Until the late nineteenth century, governments' support of science and technology tended to be erratic and sometimes capricious. Important as it was in furthering the growth of science and the scientific community, only gradually in the twentieth century did some governments begin to assume more direct responsibility for the general support of science and technology. Attention was at first directed mainly toward fields of obvious social interest, such as agriculture and medicine, but then became more eclectic—especially as the extraordinarily exciting work in nuclear physics expanded in the 1920s and 1930s. It was primarily European governments that supported science from public treasuries; in the United States, the prevailing practice of providing government funding only for well-defined practical objectives, due in part to perceived constitutional barriers, discouraged substantial public funding of what would now be called basic research. A great deal of such research was in fact carried out either with support from private foundations—a major factor in the development of science in the United States in the interwar years—or with public funds under the rubric of suitably "practical" goals such as naval navigation, weather forecasting, aircraft design, or geological surveys. Though the center of gravity of science remained in Europe in the early decades of the century, basic research developed rapidly—especially in the universities—in the United States after the 1930s, measurably helped by the influx of refugee scientists Hitler drove from Germany.

Any lingering question about the relevance of science and technology to national purposes was erased by the experience of World War II. Not only was the importance of public investment in science and technology in the security field now clear, but the potential benefits in other areas of social interest also appeared to be obvious. The spectacular applications of the theoretical studies of the atom demonstrated vividly that even basic research could be directly relevant to the achievement of national purposes. Governments, particularly those of the former combatants, took up the support of research and development as a clear national commitment. With the onset of the cold war, this commitment grew to become a substantial fraction of government expenditures, for most industrialized countries larger than the r/d expenditures of the private sector for many years after the war.

There is much variation among nations in the details of those commitments; in the United States, for example, there is to this day only limited government support for the development of technology in commercially oriented subjects outside agriculture and health. And some nations rely much more heavily on defense-oriented r/d expenditures than do others; in 1988 the United States devoted approximately 65 percent of the federal r/d budget to defense, while Japan allocated only 5 percent for that purpose.[12] The absolute amount of overall support varies widely, as might be expected, roughly proportionally to gross national product (GNP). Most of the larger Western nations, including Japan, spend between 2 and 3 percent of their GNP on r/d. The Soviet Union may have spent a larger proportion before its fragmentation, but uncertainties in the data make the comparison somewhat suspect. We will return to these figures and comparisons below, for they are an important part of the story.

The United States emerged in the postwar era as the dominant scientific and technological power, but by the end of the 1980s the overall picture was in the process of substantial change. The scientific and technological competence of many industrial nations had grown to challenge U.S. dominance, and the end of the cold war implied a decreasing role for the military and for security goals in the support of r/d. Economic competitiveness became an increasingly important issue, with the role of science and technology seen as a key factor in determining a nation's international economic position. At the same time, considerable uncertainty, especially in the United States, prevailed about what government policies were appropriate and effective in the stimulation of technological innovation. Aspects of these issues will be taken up in detail in later chapters, for they are critical to understanding the role of science and technology in the evolution of international affairs.

In sum, institutional changes in this century, particularly since World War II, have transformed the scale of the scientific and technological enterprises, and the relationship of those enterprises to a nation's economy, to its government, and to the rest of the world. Much else has changed, of course, in the role of government itself in social and economic issues, in the general wealth of a privileged portion of the world's population, in the very size of the world's population, and in the nature of the relations among nations. Applications of the results of science and technology that have emerged from these transformed enterprises have had much to do with those larger societal changes, just as those larger changes have in turn had much to do with the development of the enterprises.

The New Enterprises

To pursue our quest of understanding what this new level of commitment to science and technology means for international affairs, it is essential that we have a sense of the enterprises themselves: the scale of the resources that support them and the purposes of that support, the internal processes that determine the nature of the outcomes, the process by which goals are set, and the degree to which outcomes can be characterized or anticipated. Our purpose is not to present a complete picture of the current science and technology scene; that has been done more extensively elsewhere. Rather, it is to explore the forces that influence the science and technology system and to relate the system and its results to the evolution of international affairs.

Resources

The magnitude of the resource commitment to science and technology today has grown to rather substantial levels, both in absolute terms and as a proportion of national incomes.[13] Total worldwide funding for r/d in 1988 was roughly in the range of $450–500 billion, the overwhelming proportion expended by the industrialized nations.[14] The United States accounted for close to $140 billion, Western Europe and Canada in total about $90 billion, and Japan in excess of $50 billion.[15] The USSR's commitment that year is hard to estimate but can conservatively be assumed to have been equal to that of the United States at $140 billion.[16] The rest of the world makes up the difference. The split between private and public resources for r/d varied enormously among nations; essentially all the expenditures came from the public treasury in socialist countries, while roughly half did so in the United States and less than 20 percent in Japan. Table 1 presents some information for 1988, the latest year for which comparable figures are available, for major Organization for Economic Cooperation and Development (OECD) countries.

The ratio of a nation's r/d funding to its GNP is a reflection of its overall commitment to science and technology. Table 2 presents that information for OECD countries in 1988 (the ratios probably had not changed substantially by the early 1990s). The proportions of those resources devoted to defense and to civilian purposes vary widely among countries, with potentially important consequences for competitive position in high technology. How this balance will shift as a result of the end of the cold war is not clear at present, though a reduction in the dominance of defense in U.S. government r/d funding can be expected.

In Japan and West Germany, a strikingly larger proportion of resources is devoted to nondefense r/d than in other OECD countries. The correlation of that measure with the high-technology trade surpluses of those two countries does not prove a causal relationship, but it is certainly suggestive.

Some historical perspective is useful in considering the scale of resources now devoted to science and technology. Table 3 shows how the expenditure of government funds in the United States has grown since 1970 in current and constant (1982) dollars. The growth pattern is roughly comparable in other major nations.

The number of employed scientists and engineers in the United States doubled between 1978 and 1988, from 2.6 million to more than 5.2 million.[17] Over the same period, the number of scientists and engineers engaged in r/d went up about 40 percent in the United States, from approxi-

TABLE 1
R/D Funding in Major OECD Countries (1988)

	Total R/D Budget (billions of dollars)	Public Funds (% of total)	Private Funds (% of total)
U.S.	138	49	49
Japan	51	20	71
Germany (West)	25	34	64
France	18	50	43
U.K.	17	37	51

Source: OECD in Figures: Supplement to the OECD Observer (Organization for Economic Cooperation and Development, Paris), no. 170 (June/July 1991): 52–53.
Note: Figures are rounded.

TABLE 2
R/D Funding as a Percentage of GNP (1988)

	Total R/D Funding	Nondefense	Defense
U.S.	2.8	1.9	0.9
Japan	2.9	2.9	< 0.1
Germany (West)	2.9	2.7	0.2
France	2.3	1.8	0.5
U.K. (1986)	2.2	1.7	0.5

Source: National Science Foundation, International Science and Technology Data Update: 1991, NSF 91-309 (Washington, D.C., 1991), pp. 3–8.

TABLE 3
U.S. Government R/D Funding over Time

	Current Dollars (billions)	Constant 1982 Dollars (billions)
1970	14.9	35.6
1975	18.1	31.0
1985	52.1	46.9
1991 (est.)	66.0	48.0

Source: National Science Board, Science and Engineering Indicators, 1991, NSB 91–1 (Washington, D.C.: National Science Foundation, 1991), app., table 4–3.

mately 55 to 76 per 10,000 persons in the labor force. That proportion is higher than in any other OECD country; Japan comes the closest at just under 70 per 10,000, with West Germany next at 54.[18]

One of the surprising and little-noted facts in this general resource picture is how much r/d is supported to further international goals. The commitment of financial and human resources is obviously intended to serve greatly varying objectives; but, to take the United States as an example, a substantial portion of U.S. government support for r/d comes from agencies whose purposes are importantly, sometimes dominantly, derived from international considerations. The r/d funding of the Department of Defense, the weapons segments of the Department of Energy, some portion of the National Aeronautics and Space Administration (NASA), the Agency for International Development (AID), and parts of other agencies—more than 65 percent of the U.S. government r/d total—can be said to be justified by the expected contribution to advancing the international position and policies of the United States.[19] A considerable portion of the r/d expenditures of American industry is in turn motivated by the prospects of future contracts from those same government agencies, and by the demands of international competition in the marketplace.

Thus, a large portion of the funding for science and technology in the United States is substantially motivated by internationally oriented goals. For other countries, the picture is different in detail but not in substance. In fact, for European nations and Japan, the pressures of international competition have been more directly relevant to r/d allocations and for far longer than in the United States.

Much more than is generally recognized, therefore, international objectives have become a major justification for funding for science and technology. It can with fairness be said that the international relationships among nations have become the prime driving force of the rapid pace of technological change.

The Evolution of Technology

The majority of funding for r/d is for applied research and development—those activities determined largely by the end use contemplated rather than by the desire to advance knowledge. A rough rule of thumb for advanced countries is that in the neighborhood of 15–20 percent of total public r/d funds, and considerably less of industrial r/d funds, is for basic research.[20] Thus, on the order of 80–85 percent of the total funding for r/d (the figure is higher in less technologically advanced countries) is for applied research and development—that is, for goals for which the primary application is known in advance and for which technologies are sought with particular characteristics and capabilities.

R/d support is not the only instrument governments have for influencing the direction of technological development. Regulations, tax credits, guaranteed loans, tariff protection, and other such public-policy measures are also available and often used.

As a result, most advances in technology are a direct product of governmental and industrial processes that define what technologies are needed or wanted and that allocate resources or implement policies designed to bring them into existence. Almost all technology the world must assimilate, or cope with, is a product of calculated decisions made in existing policy processes. Technological change does *not* result from mindless evolution outside the control or influence of human decision and choice.

However, those bald statements require both qualification and clarification; they are valid but are by no means all of the explanation of how technology develops, and they can be misleading. The full story is complex, for it must recognize the many and varied elements from within the science and technology system and their interactions with social, political, and economic factors eternal to the system. Harvey Brooks suggests biological evolution as an appropriate metaphor to characterize the evolution of technology—not a perfect analogy, but one that offers valuable insights and a suggestive intellectual framework.[21]

Brooks starts by equating genetic inheritance to the inherent logic of technological development, and natural biological selection to the decision mechanisms for technological choice (prominently including the market). Just as the natural environment leads to selection in species evolution, so the socioeconomic environment, including competing technologies, determines the evolution of a given technology. Similar to selection in biological systems, technological selection is exercised through millions of decentralized and uncoordinated decisions. Conscious choice in technological evolution—by means of government regulation or public investment to produce particular technologies—is analogous to intervention in biological systems through artificial selection. In both cases, human beings have learned how to intervene in the environment so as to channel evolution in desired directions. But the ability to bring about desired change is limited by the internal logic of technological evolution, just as intervention in biological evolution is limited by the laws of genetics and by what already exists.

The metaphor also helps to clarify the differences between the technological determinists, who argue that technology proceeds entirely with an internal logic of its own, and those at the other end of the spectrum, who argue that technology is entirely a product of social forces. It is a more complicated process, as always, than simply one or the other. Brooks's comments on this are worth quoting in full:

> Just as the genetic variations on which natural selection acts are determined by internal genetic events, so evolution from one generation of technology to the next is determined by logic internal to the technological system. But just as the number of genetic variations is very large compared to the number that are propagated in the next generation as a result of natural selection, so is the number of technical possibilities very large compared with those that actually survive in the development process, and even more so in the market or society. Thus the influence of society and culture on the inner logic of technology is similar to the influence of the environment on genetic inheritance between successive generations. In each case the inheritance mechanism produces a large redundancy of possibilities, while the environment selects those that survive to the next generation. In technological evolution, what survives provides the knowledge base that generates the full range of possibilities for the next generation of technology.[22]

Thus, to a first approximation, the technologies that result from the technological enterprise are a product of human choices made at many different points of the innovation process, but always constrained by the state of knowledge at the time and by the internal logic of the technological system. The choices may be the innumerable technical decisions made during a development process; or they may be the choices made by governments in pursuit of particular research or technological objectives; or they may be the widely dispersed choices of the market made in great numbers, with little significance individually but with great cumulative influence.

It is the decentralized nature of the choices, and the fact that each is made on the basis of different and localized considerations, that gives the appearance of technological momentum out of human control. Though that impression is not an accurate reflection of reality, the tyranny of innumerable, decentralized decisions does limit how much planned control of technological development is possible.

GOVERNMENTAL AND INDUSTRIAL INFLUENCE

Science and Technology Policy

Government policies necessarily have an influence on the nature of technology that emerges within each country. Policies for science and technology, as all policies, are intended to serve the objectives of the state, which can mean many things: contributing to the overriding goal of survival; advancing the economic interests of the state, whether in competition or cooperation with other states; maximizing its citizens' welfare, defined either narrowly in economic terms or more broadly to include personal security, political freedom, and cultural attainment; or serving other values, such as preserving a particular form of government, advancing the spread of an ideology, or promoting acceptance of a religion. Or it can, and usually does, mean a mixture of many purposes—often in conflict with one another—that requires compromises and trade-offs among the various objectives.

Whatever the mix of objectives, the purpose of policy for science and technology will be directly related to those objectives. This is clearly so for applied research and development; but even basic research, not tied to specific applications, is supported largely in the expectation of long-term benefits and in the recognition of its relevance to other policy needs, such as education. At times, especially during the height of the cold war, accomplishment in basic research was also seen as an element in the political competition among states.

The role of the state in providing and influencing the nature of the support for r/d thus necessarily ties the scientific and technological enterprises to national goals and biases the processes of technological development toward those goals. This relationship is most evident in the security area, but it is relevant to others as well: space, agriculture, and health, for example, and the many other fields affected by regulatory and other policies intended to serve particular national purposes.

The national basis of decision making for development of technology has two other important consequences. First, it imparts a powerful national identification to science and technology, defining in national/domestic terms the interest of the state in their strength and productivity and implicitly underlining the element of competition in the international arena. Second, the national base binds the scientific and technological enterprises to the policy and decision processes of individual nations and thus to particular political structures, to the idiosyncrasies of national attitudes and budgets, and to the perspectives of countries acting alone rather than in a larger collectivity.

Thus, the policy processes dealing with science and technology and the enterprises that encompass them remain predominantly national, even if progress in science and technology and in their applications inexorably moves toward international and even global effects.

The patterns of evolution of technology do not necessarily cross national borders; countries tend to have rather different policy styles with regard to science and technology, resulting in disparate influences on technological development. Henry Ergas, in his comparative studies of technology policies in OECD countries, describes three generalized patterns into which most Western industrial countries fall.[23] The first he calls a "mission-oriented" strategy, characterized by a focus on a small number of technologies of particular strategic importance, with centralized decision making and commitment of resources to achieve those technologies. The United States, the United Kingdom, and France tend to share this strategy.

The second pattern is a "diffusion-oriented" strategy, characterized by decentralized policies and reliance on diffuse decision making that seek to keep up with, and take advantage of, radical innovations occurring elsewhere. Germany, Sweden, and Switzerland are primarily diffusion-oriented in their science and technology policies.

The third pattern is exemplified by Japan, which, Ergas argues, combines both of the other strategies in a unique blend that sets the country apart. It has mounted significant centralized programs to advance specific technological goals and at the same time has devoted important resources to developing a capacity for the diffusion of new technologies, largely imported from mission-oriented countries.

The concentration of support for science and technology in order to achieve technologies of major strategic importance, characteristic of mission-oriented countries, will necessarily affect the pattern of evolution of technology. This link can be seen most clearly in the national-security area, where defense objectives have had such great prominence in the funding of r/d in the United States, the United Kingdom, and France. The defense establishments of those nations tend to place primary emphasis in their weapons development and procurement on obtaining the latest in technological capabilities, at times even at the cost of effectiveness.[24] The norm is to emphasize new weapons capabilities and to anticipate all possible developments that might be accessible to others, as opposed to refining existing capabilities. As a result, maximum rate of change and technical sophistication tend to be the goals of weapons-systems development—and thus of resulting technologies.[25]

The same pattern can be observed in the support for the occasional large project outside the security area, such as Project Apollo, the American lunar-landing program; or the development of nuclear power or the supersonic airplane, in which some governments have taken on a large-scale technology-forcing role. The pattern is also evident in the civilian economy in mission-oriented countries, with emphasis placed on seeking innovative new products or processes as a way to capture commercial markets.

Thus, mission-oriented nations characteristically seek radical technological developments, both in public technology applications such as defense or space and in commercial fields hoping to base new industries on new technologies.

Diffusion-oriented countries are less likely to push technological development in radical directions than to emphasize incremental adaptation to technological change.[26] In the process, they will typically focus on factors such as cost, scope of services provided, reliability, and quality. These, it must be noted, can be critical determinants of commercial success in industries—particularly high-technology fields—in which challengers adept at incremental improvements of new technologies can create an advantage over the innovator.[27]

It is evident that a nation that can successfully combine both strategies is likely to be in a particularly strong competitive position against nations in which one or the other strategy dominates; such is the case with Japan. We will return to the issue of technological competitiveness in chapter 4.

Other Government Policies

Governmental influence on the nature of technological evolution is not restricted to policies that support science and technology directly. Regulation, setting of standards, taxation, and trade restrictions have also become powerful technology-forcing agents. Standards set by governments or agreed to in international negotiations can determine the design and economic returns of massive technological systems—for example, high-definition television or the integrated services digital network (which are discussed in chapter 4). Pollution regulations, such as those requiring reduced auto emissions or lower sulfur emissions in the burning of fossil fuels, can call forth new technologies to meet those mandated needs. And the growth of and broadened interpretation of product- and environmental-liability rules can stimulate the development of both "safe" technologies and the means to anticipate undesirable technological externalities.[28]

Thus, while these other policies may not move the research front of technology (the knowledge base, in Brooks's evolutionary metaphor) as fast as policies targeted directly on the development of technology, they can have important economic and political consequences both domestically and internationally. Moreover, the role of regulation and standard-setting in technological development is likely to grow in scope and importance in years to come as the uses and side effects of technology create public pressures for government action.

Industrial Incentives

For industry in market economies, the prospect of profits in the marketplace is the primary influence on the funding of r/d. As a result, the dominant forces affecting the evolution of commercial technology would presumably be best characterized as "market-pull" rather than "technology-push," with decentralized decisions by consumers creating the market. Since the market cannot be aware of new technological possibilities before they are introduced, it could be expected that gradual rather than rapid evolution would be the norm. It is a much more complicated process than that, however, and one that varies among nations.[29] Even so, there are several generalizations that are important for their bearing on the character of emerging technology.

Nathan Rosenberg identifies three primary historical inducements to technological change, all in the context of expectation of profit: correcting imbalances in technology (improving one part of a system to compensate for obvious and costly limitations in another); avoiding vulnerability to labor disruption (substituting technology for labor); and protecting against disaster (insulating against a cutoff of supplies or catastrophic technological failure).[30] I would add three other, more recent, inducements that arise from the changed makeup of the incentives and rewards that affect the scientific and technological enterprises.

The first stems from the growing role, just described, of governmental regulation, taxation, and related policies that directly or indirectly affect technology. Barely existing before this century, that influence on technological evolution is now substantial, and it is bound to grow.

A second inducement is the expectation of substantial government procurement, particularly in the defense area. That expectation has led industry to commit r/d funds and resources to developments that correspond to the objectives of the defense establishment—that is, to push technological capabilities ahead in relevant areas as far and as fast as possible.[31] The ever-closer marriage of science and technology within industry contributes to that process, by more rapidly advancing the state of knowledge that makes further technological evolution possible.

The third inducement is likely to be the most significant for the pace of technological evolution, and for industrial structure and international competitiveness as well. It is the emphasis on innovation that now characterizes high-technology industry, in the context of a marketplace that provides large rewards for rapid change in high-technology products. In fact, the dominant corporate strategy for those industries most involved in high technology—computers and communications in particular—can be described as maximizing the rate of innovation in a competitive market environment.[32] Technological change, in that case, becomes the norm, the primary goal of a company's r/d. It can be a particularly disruptive quest, for at times r/d will result in new capabilities that will compete with existing product lines rather than reinforcing them; in such cases a firm will face agonizing choices and often the inability to respond adequately to the new competitive environment.[33] It can be even more disruptive if, as is often the case, new high-technology products are developed by companies that were not engaged in the predecessor market; the successful leaders in transistors, for example, did not turn out to be the companies that had dominated the displaced vacuum-tube market.

Thus, there is a new emphasis in some industrial sectors on the development of technology before the identification of need—a technology-push factor in the evolution of technology that encourages innovation as a primary goal, with utility, cost, and other decision elements considered only after the outlines of the technological possibilities become clear. There have been examples in the past of the development of new technologies that had little relevance to a perceived need. Until recently, however, most were not deployed—they simply became the knowledge base for later innovations or were forgotten. But this is surely the first era in history in which there is heavy investment in the development of new technology with little knowledge at the outset of what its ultimate market value will be.

It is interesting to note that the rapid commercial development of technology due to these factors has now generally outpaced military-sponsored development, creating a flow of technology from the commercial to the military sector. That is a reversal of the "spin-off" experience of the postwar years, and it has led to increasing problems with the possible dual use (commercial and military) of new technologies.[34] It has also meant that Japan, which devotes a very small portion of its r/d expenditures to defense (less than 4 percent versus the United States' roughly 65 percent), now finds itself in the position of having advanced technologies developed for the marketplace that not only support a commercial lead, but are also of growing relevance to possible military application.

Other goals, of course, also influence private-sector interest in technology. To survive, all industries, especially those in a highly competitive environment, must be concerned with factors such as cost, scope of services provided, reliability, quality, and, increasingly, environmental effects. This is true even of industry in mission-oriented economies, though in those such factors do not typically receive as much attention as in economies dominated by a diffusion-oriented strategy. A substantial portion of private sector r/d—a portion not easily measured—is devoted to improving or expanding cost and quality factors in a company's product mix. The technological results of that r/d are likely to be much more incremental in nature than are those of the r/d devoted to bringing forth new technologies. The r/d aimed at incremental improvement rather than creation of new technologies may well be the determining factor in economic performance, however, even for the innovating companies. Challengers adept at rapid reproduction and modifications of new products can severely reduce the market share the innovator is able to capture, unless the innovator is able to stay ahead through steady improvement of the original product. The success of Japanese companies in the importation and improvement of technology developed first in other countries demonstrates well the viability of this strategy.[35]

It should be noted that in the countries with command economies, the influences on the setting of technological goals and the expenditure of r/d funds as a whole are dominated by government policy, and in that respect the role of government is parallel to that in market economies. However, the generally lagging scientific and technological capacity in command economies—the reasons for which will be explored in chapter 4—has tended to mean they have primarily been users of technology developed in market economies, and much less commonly have been serious contributors to technological evolution.

INFLUENCE FROM WITHIN SCIENCE AND TECHNOLOGY

Many factors within the scientific and technological enterprises also affect the evolutionary process and thus the character of the technologies that emerge. Some factors have always been present; some are new or greatly altered as a result of the profound changes in the setting and role of science and technology.

The Limits of Technological Optimism

The spectacular advances in technology in recent years, and the much closer coupling of science and technology, have contributed to the uncritical view that science and technology offer limitless means for advancing human welfare and coping with the world's problems. Notwithstanding the companion view that sees a dark side to technology—that sees it as the source of many problems—this general optimism about the ability of technology to help overcome difficulties or meet desires has been an important element in the growth of resources for r/d intended to achieve specific public and private objectives.

This optimism at times leads to the view that there are few constraints on technological possibilities, that technologies can be designed to solve all problems or reach all goals, if only enough financial and human resources are allocated. New "Manhattan Projects" are often proposed as cures for planetary or human maladies.

The reality, of course, is rather different. To hark back to the evolutionary metaphor, possible technologies are always conditioned by the state of knowledge and by the internal logic of technological evolution, just as intervention in biological evolution is conditioned by existing life forms and by the laws of genetics. The knowledge base is undoubtedly expanding faster, and across a wider spectrum, than ever before, contributing greatly to that sense that all problems are soluble. But there are important constraints imposed by the internal structure of the system.

At the fundamental level, some technological capabilities are simply barred by physical principles—communication at greater than the speed of light, for example. Wishful thinking about overcoming underlying laws of nature may be common, but is unavailing. Surprises in scientific research are always possible—in fact, inevitable—but technology cannot be planned on the assumption that new scientific knowledge will disprove existing physical laws. Even if fundamental principles did not stand in the way, some desired technologies might require materials, knowledge, or calculations that are not available with the present knowledge base and tools; and we cannot assume they will be available when needed.

On more subtle grounds, many technological goals that involve the performance of complex tasks require a degree of systems integration that proves to be difficult or unrealizable in practice and whose accomplishment cannot be assured in advance. The more demanding the requirement for such systems integration, the greater will be the uncertainty and the lower the probability that the system will function as desired. The Strategic Defense Initiative (SDI) as originally proposed (something equivalent to an astrodome over the United States, impervious to missiles) was a rather extreme example of such a technological objective; it would have had, if it had ever been built, unprecedented requirements for systems integration (see chapter 3 for a detailed discussion of SDI).

Paradoxically, excessive optimism about the short-run possibilities of a new technology often coexists with conservatism and lack of vision about the long run. When the transistor was developed, for example, it was widely touted as a greatly superior replacement for the vacuum tube; there was no inkling of the much more astounding capabilities of integrated circuits, which transistors were to make possible. It is a vivid illustration of the limits of human imagination in attempting to anticipate the effects of new ideas within the constraints of existing experience and capabilities.

Movement of Information

A second important factor within the science and technology system affecting technological evolution, one with particular relevance for international competition, is the pattern of movement of scientific and technological information. The pattern is not fundamentally different from what it was in the past, but the scale of the scientific and technological enterprises, the spectacular advances in communication and transportation technologies, and the development of large, interactive, worldwide communities of scientists have greatly increased the volume of transfer that takes place, the cross-linkages among fields, and the interaction of technologies with one another.[36]

Scientific information generally moves through open channels; unimpeded communication of results is essential for the effective cumulation of knowledge. The principles of openness, considered to be essential for a healthy scientific enterprise, are reflected in the values and norms of the scientific community, which date from its Baconian origins and which have been embedded in its structures of professional societies and publications. There have been attempts at various times to restrict the flow of scientific information, especially for national-security reasons during the cold war. Such restrictions tend to be only marginally effective and are usually counterproductive both for security and for science.[37] Attempts at restriction are likely to emerge again, however, as technologies with significant economic potential become increasingly science-intensive. Signs of such attempts are evident in advocacy of restrictions on the access of foreign scientists to research at American universities or to potentially strategic areas of basic research, such as semiconductors, molecular biology, or superconductivity.[38]

Information about technology, by comparison with scientific information, is more susceptible to control; considerable effectiveness is possible in limiting its availability for a period of time. Most companies depend on that time advantage for their ability to secure a profit from their technological developments. Eventually, however, there can be few secrets even in technology: others duplicate the work, the deployment of the technology betrays vital elements, later advances lessen the importance of earlier work, secrets are leaked intentionally or through covert action, or in some cases the mere knowledge that something has been done is enough to make replication possible. The more radical and fundamental an innovation, the harder it is to prevent others from finding out about it and reproducing it, independently if necessary. In fact, it is the incremental innovation on fundamental concepts that is often easier to protect from competitors than the concepts themselves because it involves more implicit "know-how," which is a product of experience rather than codified technological development.[39]

Thus, the essential information for widespread technological evolution is ultimately available to all who seek it and who have the resources and competence to make use of it. This does not deter nations from approaching technology as a national, even nationalistic, enterprise, as we noted before. In practice, short-term technological advantage can convey important benefits to the state, especially economic benefits in fast-moving high-technology areas. Over time, technological advantage becomes harder to maintain as knowledge spreads and as technological competence irresistibly develops in other nations. Leadership can be maintained, but it must be done through accomplishment rather than through ultimately futile attempts to deny knowledge to others.

The inevitability of the flow of information has other implications, one of the more important being that it is essentially impossible to prevent the development somewhere, sometime, of feasible technologies for which there is strong motivation—even if they are thought to be politically or militarily undesirable. If such technologies require scarce resources (such as fissionable material) or are particularly costly, it may be possible at least to delay their development; but eventually they cannot be prevented if there is sufficient political or economic motivation and the resources are available.[40] Ironically, attempts to limit deployment of some technologies—for example, through arms-control agreements—can increase the incentives for the development of related technologies that are not excluded by the agreement and that in the end may devalue the agreement.

Synergism among Technologies

That irony is related to a third factor of increasing importance in the science and technology system: unplanned and unexpected synergistic effects among technologies. Synergism has always been a characteristic of the evolution of technology, but the now much larger scale of the scientific and technological enterprises, the dispersed nature of decisions about technology, and the rapid diffusion of information all combine to produce unexpected and at times highly significant new technological opportunities. Development efforts aimed at increasing the sensitivity and performance of radio telescopes, for example, had major applications in commercial high-fidelity equipment; advances in solid-state electronics for commercial purposes made possible equipment miniaturization that in turn greatly enhanced the feasibility as well as the capability of ballistic missiles. These cases illustrate a general principle that new technologies, whatever their source, often will have applications far from the original purposes for which they were developed. This principle adds to the element of surprise and unpredictability in the evolution of technology, its applications, and its externalities.

Unexpected applications of technology in scientific research may be equally significant for the advancement of science. The field of radio astronomy, for example, was created when greatly improved radio receivers developed for ground-based application unexpectedly picked up signals from space outside the visible portion of the electromagnetic spectrum.

The Path from Fundamental Knowledge to Application

The apparently closer relationship than in the past between the results of the research laboratory and potential technological applications, however, makes it more likely that the pattern of support for basic science will now be an important influence on the direction of technological evolution. It is clear, for example, that the explosive progress in molecular biology will, over time, have major implications for the development of technologies for control of disease and increased agricultural productivity; that advances in the physics of materials, such as those related to superconductivity, will one day have a large economic impact through energy-technology applications; and that developments in pure mathematics will continue to be applied in the design of computers.

The exact nature of the discoveries, or how or when they will result in usable applications, cannot be foreseen. But it is now more likely to be true that advances in some fields of science will find a relatively direct route to their application in technology, sometimes with large and rapid financial implications. One notable effect, among many, is to encourage more investment, and ultimately a faster pace of change in "hot" scientific fields that are expected to have significant economic consequences. Another is to elevate basic research to a more prominent place in relation to economic goals and international competitiveness than it occupied in the past.

In addition, as many technologies have become more science-dependent—that is, embodying proportionally more direct scientific knowledge—the relation between laboratory and application has necessarily also become closer. This, too, serves to increase the perceived relevance of basic research to economic and international goals.

This change in the relationship of fundamental knowledge to technology does not mean that scientific surprise is any less likely. Surprise will come not only from the unexpected synergism of new knowledge from different sources or fields, but more fundamentally from the fact that new knowledge in science is unpredictable in detail, and at times even conceptually. Scientific theory attempts to build upon known facts and accepted ideas to hypothesize yet-unknown or unobserved relationships or phenomena. But those hypotheses cannot take account of that which has not yet been imagined, so that prediction and actuality have often turned out to be strikingly different. The revolution in physics in this century that developed entirely new concepts of the quantum structure of matter was not foreseen by scientists of earlier eras. Yet it was that knowledge that made possible many of the technologies we are grappling with today—including, most vividly, nuclear weapons.

The discovery in 1986 of superconductivity at higher temperatures than previously thought possible, with the physics of the phenomenon still incompletely understood, was another example of the inability to anticipate new scientific knowledge.[41] The discovery galvanized the scientific community in that and neighboring fields; it opened new questions for exploration and understanding and may well trigger an expansion of knowledge over a much broader area. It will also, quite certainly, produce economic advantage for those able to translate the knowledge into successful technological application.

Moreover, nature is complex, so that outcomes of known processes are often not predictable. The number of neutrons emitted in the fission of uranium, for example, is dependent on the configuration of the material and the characteristics of the surrounding environment that might impede or capture released neutrons; the total number actually emitted under bombardment could be different without changing the basic theoretical model of the nucleus.[42] If it were not possible to configure the interaction so as to release sufficient neutrons of the appropriate energy to sustain a nuclear chain reaction, uranium-based nuclear weapons would not have been possible, though the rest of nuclear physics would have been little affected.

Thus, notwithstanding the steadily closer relationship between the laboratory and technological application, there is no way to anticipate all or even most advances in scientific knowledge; surprise is inevitable. Even if in retrospect it seems that some unexpected scientific knowledge might have been foreseen, in practice all developments that are judged in hindsight to have been "foreseeable" could not actually have been selected from among the many possible "futures." Human imagination is not up to the task.

Pace of Change

Considering all the expansionary elements in the scientific and technological enterprises, it would seem to be a reasonable proposition that the rate of change in science and technology and their applications is steadily accelerating and that, as a result, society will have to cope with an ever-increasing flood of technology-induced innovation. Qualitative experience would seem to support the impression; rhetoric usually assumes it to be true.[43] But the proposition is not as certain as it appears. The time lag between the laboratory and commercial application of new discoveries is not necessarily different now from what it was in previous centuries. The transistor, laser, and computer all took decades to have an appreciable impact on industry, while some earlier inventions—such as the telephone in the nineteenth century—were widely and quickly adopted.[44] Developments in biotechnology have also been slow to reach the marketplace, and it is not at all clear that the newly discovered higher-temperature superconductors will find their way into widespread commercial use any more rapidly. The increase in the speed of communication during the nineteenth century—in 1800 information could travel no faster than a horse and by 1900 had reached the speed of light—was actually proportionally much greater than increases in the twentieth century.[45]

There are many definitions involved in these assertions that would need careful delineation if this issue were to be probed in detail (for example, should the commercialization of the telephone be measured from Faraday's work on electricity or from Bell's invention of the device? is the speed of communication or the capacity of communication links now the more important variable?). In fact, as Rosenberg argues, the lag between demonstration of technical feasibility and commercialization is a function of both technical and economic factors.[46] That is, the time it takes for a new technology to emerge on the market depends on the need for related technical advances (e.g., acceptable production techniques) and on the economics of the marketplace (e.g., the relative cost of alternative products already available). There is no consistent time lag applicable to all inventions.[47]

Whether or not change is accelerating is an interesting intellectual question, but one that is not central to most of our inquiry. It is enough to note that this widespread sense of accelerating change due to science and technology does not necessarily mean there have been actual reductions in the time it takes to move from knowledge to application, or more-rapid changes in technological capabilities.

The justification for the proposition of accelerating change, and for the common acceptance of its validity, lies elsewhere. One contributing factor is the relatively steady-state nature of humankind's experience before the scientific revolution; the rapid progress in understanding the physical world and in the development of technology has been a product only of recent years. Another factor is the breadth and the scale of today's scientific and technological enterprises, as opposed to the pace of change within any given segment. Advances come across a much broader front than in previous eras, with correspondingly greater opportunities for synergism within science and technology and with much more extensive and complex interactions with society. And finally, the increase in scale of many other attributes of society—population and wealth in particular—creates, when coupled with new developments in science and technology, more evident and more widespread changes that impinge on human consciousness.

There is good reason, therefore, for the sense of accelerating change; but its validity rests on the choice of variables of interest, rather than on a clear and demonstrable phenomenon.

Social Sciences and Social Technologies

Though we indicated earlier that the social sciences and social technologies will not be a major focus of this work, it is important to note that they are also factors within science and technology that affect the evolution of technology.

The formal organization of the social sciences along disciplinary lines began in the late nineteenth century, though some fields and applications are much older: the concept of a census—basically a social survey—was built into the American constitution, and rigorous economic analysis based on available data dates from the eighteenth century.[48] The fields have, of course, grown and multiplied enormously, especially in the United States, receiving support from both government and industry for research and policy advice.[49]

Many of the social sciences are now heavily engaged with subjects that stem in some measure from issues posed by, or stimulated by, technological innovation. Studies of behavioral responses to stress in the environment, toxic-chemical risks, military strategy, effects of population growth, energy policy, and technology policy—all are examples of social-science research interests that are tied in varying degrees to the societal effects of technological innovation. The ability to study these issues reproducibly and systematically, with clear rules, itself constitutes a technological innovation that has been increasing in importance and application.

Today, the growing social-scientific knowledge about the scientific and technological enterprises themselves, and about the management of innovation, has become an important element in the evolution of the scientific and technological system itself. Much of that social-science research, in fact, is motivated by the need to understand the scientific and technological system and its interaction with society, so as to make it better able to serve social goals. We have drawn on that research extensively in describing the history and process of technological development, and we necessarily will continue to do so throughout the succeeding chapters.

Patterns of Outcomes and Effects

The science and technology system that has evolved is one that is dedicated to stimulating advances in technology so as to expand capabilities and performance. The incentives and the structure that make the system work in this way are now institutionalized in the scientific and technological enterprises themselves and in their economic and social setting. The result is a stream of technological outcomes that contribute to a steady, sometimes spectacular growth in the capability to carry out new tasks and to perform existing tasks at a faster rate, at a greater (or much smaller) distance, with greater precision, with higher quality, with greater efficiency, with fewer people, at lower cost, with more power, and with other enhanced characteristics.

It is useful for our purposes to identify, even at high levels of generality, some of the first-order societal effects that will flow from these technological outcomes. There are important pitfalls in making such generalizations—and in prediction of social consequences at all. In particular, it is evident that technological change is not uniform across fields, nor is it independent of policy, so that disaggregation by fields or by differences among nations can result in significant differences in the nature of technological developments and in their eventual implications. Nevertheless, the general patterns identified below serve to highlight the implications of technology that will continually recur in our more detailed exploration in succeeding chapters.

Broader Opportunities, Choice, and Flexibility

A consequence of expanding capabilities is that the horizons of choice are broadened for individuals, organizations, industries, and governments. The promises that science and technology offer—to cure disease, to solve pressing problems, to increase availability of necessities or develop substitutes for them, to provide new services (often on a global scale), or to achieve many other varied and dramatic goals—create an increasing array of options for policy in all societies. Inherent in that new flexibility is the greater responsibility placed on social decision mechanisms for the wisdom of the choices made.

Growing Internationalization and Globalization

Two effects are at work here. The first is a result of the development of technology that expands capabilities of size, distance, and power, thus becoming increasingly international in its reach—either because it must be deployed in an international setting (such as space systems) or because its effects have an unavoidable international impact (such as atomic weapons or information technologies).

A second effect results from the cumulative externalities of countless small decisions affecting technology that integrate over time and space to create implications that extend beyond national borders, and increasingly to the planet itself. The widespread diffusion of television is an example. Concerns over global warming and destruction of the ozone layer, both stemming from the effluents of widely used and essential technologies, represent obvious examples of another kind.

A critical consequence of both effects is the much more rapid and effective diffusion of information on a global basis, as well as increased difficulty for governments in preventing the diffusion of knowledge of events or of ideas.

Growth of Large Technological Systems

As capabilities expand, it becomes possible and efficient, often necessary, to link technologies in ever-larger systems that are now increasingly international in scope. Once the systems are in place, however, the sunk costs and fixed installations tend to reduce flexibility, since change can come only slowly and at substantial cost. The global energy system, now so wedded to fossil fuels, provides an apt and disheartening illustration of a massive system with major environmental implications that can be altered only gradually over many years.

Interlocking of Societies and Economies

The result of expanding technological systems and internationalization of technology is the growing interlocking of national economies and societies. This phenomenon, commonly called interdependence, is one of the most widely remarked international effects of technology. Examples are endless, and so is the rhetoric.

Alteration of Factor Endowments

Technological change, comprising both new systems and the evolution of existing technology, means corresponding changes in the factor inputs that determine, *inter alia*, the costs of manufacturing processes or the demand for raw materials and energy. Indeed, one of the characteristics of continued technological development is that it brings down, often dramatically, the costs of the inputs required for a particular function, as r/d leads to expanded output per unit of input. Technological advance that alters factor costs is not a new phenomenon, but the continuousness and breadth of change in technology introduces a fluidity to factor costs, and hence to factor endowments, that in turn means continuing change in the economic potential of states and in their relative comparative advantages.

Diffusion of Physical Power and Capability

The technological developments that make possible high-yield explosive power in small packages able to be delivered with accuracy at a distance and at low effective cost substantially lower the barrier to the diffusion of physical power to nations, to insurgents, to terrorists, or even to individuals. The inevitable spread of technological information over time makes the effect essentially irresistible. Actual or prospective nuclear proliferation is an obvious example; the devastating use of shoulder-fired Stinger antiaircraft missiles by Afghan rebels and the acquisition of Scud missiles by Iraq are at least as striking, for they dramatize how widespread is the availability of new high-performance, cost-effective, technological weapons systems that have significant military effects.

Physical power need not imply only military uses; the ability of nations or corporations or individuals to carry out activities on a scale that has direct and substantial effects on other nations, or on global conditions, is also growing. Examples that illustrate the point are numerous: the plans (now suspended) for reversing the flow of Arctic rivers in the Soviet

Union, the destruction of tropical rain forests, and the appearance of computer "viruses," written by a single programmer, that can cause whole systems to malfunction.

Increased Science-Dependence of Technology

The closer relation of technology to the products of the basic-research laboratory implies, *inter alia*, that the health of science and the results of fundamental research have greater significance for technological progress and economic competitiveness of nations than at any time in the past.

Increased Complexity and Uncertainty

Synergisms, the increased sophistication of science and technology, and the inevitable uncertainties in development processes all imply that technological outcomes and their interactions with society are certain to be complex and impossible to foresee in detail.

Discovery of Problems and Causes

Important outcomes result not only from the development and application of new technology, but from the design of analytical methodologies for probing complex issues, from the expanded research to understand causes of change, and from the ability to construct complex models to forecast future conditions. These capabilities, along with more-powerful tools for experiment and computation, will continue to find problems that had not been realized, challenge desirable goals by detailing undesirable consequences, point out the need for potentially costly measures to deal with issues not yet physically demonstrable, and complicate public issues with information not readily understandable to the public or to those who must make policy decisions.

Permanence of Change

The institutionalization of scientific and technological development in government and industry guarantees that the technological environment, and thus the social environment, will never be static. New capabilities, new opportunities, altered competitive balances, unexpected problems, and fresh surprises will continue to lead to changes in the structure of societies and of issues, including, not least of all, international affairs.

Increased Significance of Science and Technology

Finally, it is worth recording the obvious: that the manifold interactions of science and technology with the social system have had the effect of increasing their importance in the functioning of society and in the policy processes concerned with their governance. Science and technology have become part of almost all social issues—sometimes only a minor part, sometimes a part with central importance. They are significant factors relevant to the making of policy in a great many areas, not the least being the international relationships with which we are primarily concerned.

These first-order effects of the scientific and technological enterprises will continue as long as science and technology are supported by society in roughly the manner and on the scale that they are today, and as long as they are injected into an economic and social system similar in fundamentals to that of today.

A particularly interesting question is whether there are consistent biases in the larger, second-order effects on society that come about when these first-order outcomes interact with social variables. Are there predictable trends that favor certain directions over others in the evolution of society—for example, stability over instability, open societies over closed societies, fragmentation of power over concentration of power, or homogeneity over diversity?

It would be satisfying to be able to offer clear and simple judgments on those alternative possibilities, for that would provide definitive propositions for assessing the future international implications of technology. Such conclusions are not possible, however, at least in a study of this kind. For every dichotomy of possible effects, it is possible to construct abstract arguments that show how the actual implementation of technology could in practice support either position; decisive proof of a dominant bias in technology's effects would be difficult or impossible to obtain.

As we will see, however, some judgments will be possible that show that inherent biases in the influence of some technologies do in fact exist; in other cases no such consistent bias will be demonstrable. These judgments will become evident as we proceed with the substantive analyses in the succeeding chapters.

PART TWO

WITH THIS BACKGROUND of the nature of the scientific and technological enterprises and their outcomes, we can now turn to an exploration of the evolution of international affairs as it is affected by, and interacts with, those enterprises and outcomes. To provide a framework, I have chosen to separate the subject into four categories: three that divide the universe of issues among national security, economics and politics, and new issues of global danger; and a fourth, crosscutting subject, governance. Though its elements are clearly not parallel or mutually exclusive, this structure offers the most useful disaggregation of the broad and ill-bounded domain of international affairs, in which technological change has so intertwined issues as to make it difficult to consider any without reference to a wide variety of others.

National-security issues, even if they no longer consistently dominate the international agenda as they did during the cold war, remain the most obvious example of international matters dramatically affected by science and technology. The importance of security issues during the postwar years and their effects on the structure of the international system would justify careful examination of the role of science and technology in any case. Moreover, even with the reduced danger of conflict among major nuclear-armed states, the possibility of armed confrontations at all levels has by no means abated and may very well increase. National security remains a primary responsibility of government, and will continue to be strongly influenced by, and in turn will influence, the march of science and technology.

The second category will consider those many issues in economic and political realms outside the traditional definition of security that have been influenced by technological change and that are, or have become, important elements of international affairs. Some issues have taken on new significance because of the changes on the world scene at the end of the cold war; some, in fact, had a hand in bringing those changes about. Though this category presents a wide array of subjects, the common patterns among them dictate that they be considered in close proximity.

Among the emerging issues that we have called global dangers, those of particular interest for our inquiry have been separated for special attention. These are the issues that may pose more direct challenges to the existing structure of the international system—climate change is the most obvious example—than any other subject on the international agenda. One of the central questions posed early in this study is whether the system as presently constituted can meet the needs for collective action that might arise if the entire planet is shown to be in jeopardy. To throw light on that question will require detailed appreciation of just what these new dangers are and of the nature of the political challenges they pose.

Finally, a chapter on the institutions and processes of governance will build on the preceding analysis to show how central aspects of governance at the national and international levels have been altered and challenged under the influence of technological change.

Three

National Security

NATIONAL SECURITY is a concept rather less well defined than the unwary would expect. It clearly encompasses as a minimum the responsibility of a government to protect its citizens and territory and to ensure the nation's survival, a responsibility typically assumed to lie primarily with a nation's military forces. In fact, that assumption of narrow dependence on the military is now quite a constrained view, as it becomes increasingly evident that the true security of a state depends on much more than military capability alone. Economic strength, competitive industry, educational quality, a wholesome environment, welfare of the citizenry, adequate public health, and effective leadership are increasingly recognized to be essential elements of a nation's real security. The fading of the apocalyptic confrontation between the superpowers further reinforces this view.

However, to avoid the danger of losing structure in the analysis by examining all issues at once, we will consider national security primarily in its military dimensions, leaving other dimensions to subsequent chapters. This focus is essential for our inquiry in any event, for military power is both a key element of international affairs and inextricably tied to progress in science and technology.

Military power has always played a central role in the relations among peoples and states; both statesmen and theorists have traditionally seen it as the primary ordering principle in the international political system.[1] Elements that affect that power are thus necessarily important to international affairs; science and technology have long been of particular significance, as a technological advantage in weapons or in support capability often led to decisive superiority on the battlefield. History is replete with examples of the use of new technology that resulted in dramatic victories and to turning points in the life of nations and empires: the development of the stirrup in the Middle Ages, use of the English longbow at Crécy, introduction of all-weather sailing vessels in the thirteenth and fourteenth centuries, the superior transportation and communications of the English and French in the Crimea, development of radar in the 1930s and 1940s, and, of course, the atomic bomb in World War II.[2] Superior technology alone by no means guarantees its effective use or actual military superiority. The introduction of machine guns in World War I heavily favored the defense against massed charges, but military leaders refused to accept the overwhelming tragic evidence of the battlefield. In 1870 the French had a rifle superior to that of the Prussians, but were defeated by superior generalship.[3]

The twentieth century has seen the most extensive changes in armaments and concepts of their use, particularly from World War II onward as the new relationship between the military and the scientific and technological enterprises came into being.[4] The spectacular developments of this century—for example, the atomic bomb, intercontinental missiles, and the computer—can blind us to how much change occurred in the nineteenth century, as nations went from military reliance on animal and wind power to steam and gasoline engines, from hand-carried messages to the telephone, and from muskets to machine guns. But it is the creation of the institutional system for stimulating technological innovation and change, together with those more recent developments, that is so different from what came before.

Today, the pressure to stretch all parameters (as was discussed in chapter 2), whether they be size, speed, power, or capacity, is a natural result of the close linkage between the scientific and technological systems and military goals. The military rewards for extending limits are seen to be substantial, and thus effective in stimulating innovation. To caricature the decision process, if some information is good, more must be better; if a thirty-minute delivery time is good, fifteen minutes must be better; if one-meter resolution is good, one-half meter must be better; if one warhead per missile is good, ten per missile must be better. Sometimes more *is* better, but the choices made during the development of technology rarely ask whether it is or not. The decision for actual deployment of a new technology might consider the question; but given the pervasiveness of the military commitment to achieving technical superiority, it is tempting for a nation to want to take advantage of every apparent advance.

This tendency for extending all parameters is fueled as well by a predisposition in security matters to use "worst-case" scenarios as a way of defining needed capabilities. When preserving a nation's security is seen as the ultimate responsibility of government, it is a seemingly irrefutable argument that armed forces should be designed to cover all possible eventualities. Moreover, since the intentions of possible adversaries cannot be definitively known, estimates of the capabilities they *could* have seem to be "harder" data, and thus carry more weight in policy deliberations, than fuzzy estimates of intentions.

The result of this logic since World War II has been the deployment of massive weapons systems in the cold-war confrontation between East and West, of a scale and character that far exceed what could have been predicted at the outset of the arms race, or what is needed for these nations' actual security.

But has the development of these massive systems modified the role of military force in international affairs? How different is it now from the past? One simple but crucial observation is that the effectiveness and relative strength of a nation's military forces are now a function of technology—of various aspects of the development, quality, and deployment of technology—more than at any time in the past. Some technology-related changes have more-profound implications, modifying permanently the dynamics of military competition and altering traditional concepts of the purpose and use of force. It is these larger changes, which have correspondingly the largest effects on international affairs, that we will take up first. We will subsequently explore the broader canvas, in which technology-related change has modified, and is continuing to modify, the many other dimensions and implications of military force in interstate relations.

Nuclear-Weapons Systems

The most consequential change in military forces in this century is, clearly, the introduction of nuclear weapons, and particularly the coupling of fusion (hydrogen) bombs with intercontinental ballistic missiles. These weapons possess overwhelming destructive potential against which no defense is in sight, or is likely ever to be possible—giving offensive strategic forces effective dominance over the defense. They represent a dramatic example of the result of the incentives inherent in the military/technological system to expand weapons capabilities in dimensions of power, reach, speed, precision, and mobility.

The most startling effect of the existence of these forces is that they cannot be used in actual warfare against a nuclear-armed enemy. That, of course, is not what the operational plans say, for both the United States and the USSR responded along familiar lines to the threat posed by the other's weapons: only military strength could balance the weapons of the other and deter their use; only the threat of swift and assured retribution on a massive scale, and thus requiring similar weapons, could provide a credible deterrent to their initial use. But it has become evident that the combat use of these weapons in an attempt to achieve some national purpose puts at immediate risk the existence of the nation as a viable entity. War, at least war involving or threatening to involve nuclear weapons on both sides, can no longer be seen as a rational policy option.[5]

Fear of the consequences of new weapons technology is not a novel reaction: the crossbow was actually banned at the Second Lateran Council in 1139 for being "too lethal"; chemical weapons were proscribed by the

Geneva Protocols of 1925; recent agreements have banned biological weapons; and other weapons have been either banned or touted as making war "obsolete."[6] But nuclear weapons genuinely threaten apocalypse, as the others do not. Early attempts at international control failed; instead, the United States and the USSR engaged in a macabre competition to prevent the other from gaining superiority and to deter use of the weapons.[7] The result was continued development of nuclear weapons and associated systems, so that by 1991 the United States and the USSR each had an arsenal in excess of ten thousand separately targetable strategic nuclear warheads, with elaborate operational plans for their use.[8]

The literature on the often-arcane disputes associated with nuclear weapons, deterrence, and their surrounding systems and doctrine is undoubtedly more extensive than that on any other aspect of postwar military affairs, or international affairs for that matter.[9] Gradually, and always with some dissent, publics and leaders on both sides have come to accept the fact, and acknowledge it openly, that these weapons cannot be used in actual warfare against a nuclear opponent.[10]

The major political effect of this situation of mutual nuclear deterrence was to impose a level of stability on the relations among nuclear-armed states that has made wars among them essentially unthinkable. It is quite likely that without nuclear weapons, the United States and the USSR would have come to blows at some point after 1945. But the danger of the use of nuclear weapons made the two countries extremely wary of any military encounter between their respective forces (though not between the forces of client states), lest it get out of control.

In fact, the nuclear powers proved to be constrained as well in their use of nuclear weapons against third-world countries—even to the point of accepting defeat, as in Vietnam and Afghanistan. In part, this restraint was due to concern about the possibility of triggering the entry of a nuclear state by attacking its client; in part, it was simply a recognition that the magnitude of destruction and the political repercussions that would result were effective deterrents to the use of these weapons, even without the fear of nuclear retaliation. The unquestioned superior power accorded by nuclear weapons has not in the event proven able to be translated into use in actual conflict situations.

It should be noted that there has been no overt change in the relationship of military power to national security. The nuclear portion of military power, however, exists only to deter initiation of conflict, not for its military potential in conflict (though some continue to assume that a presumption of war-fighting capability is a necessary part of the deterrent). Even when nuclear weapons are seen by small states as necessary to redress the conventional-force superiority of potential adversaries (Israel's situation is an illustration), the essential role of the weapons is deterrence, for actual use of the weapons if conflict erupted would almost certainly not save the state or its citizens. Moreover, the acquisition of nuclear weapons by a small state is eventually self-defeating; it becomes an incentive for similar acquisition by competitors.

To put it succinctly, when nuclear weapons are involved on both sides of a confrontation, military force can fulfill its traditional function of protection of the state and its citizens only as a deterrent, not if actual conflict ensues. Even in a confrontation in which only one side has nuclear weapons, they are largely irrelevant to the projection of power, and of highly questionable and transient value as a deterrent for small states.

It was apparently his realization that military forces could not physically protect the state in the event of hostilities (though the justification was couched in different terms) that led President Reagan to launch the Strategic Defense Initiative (SDI) in 1983 as an attempt to develop a defense against nuclear-armed ballistic missiles.[11]

The nature of the nuclear relationship between the two superpowers also served to make them uniquely dependent on decisions of the other for their very survival. Mutual dependence is, of course, inherent in all international relationships that are not totally dominated by one side. But a dependence that stems from the possibility of a unilateral decision to precipitate mutual annihilation is surely without precedent.

Science and technology, it is important to recognize, did not *make* any of these things happen. The developments emerging from the scientific and technological systems made them possible, confronting nations with new choices with greatly altered conditions and payoffs. But it was how those developments were used and how the direction and pace of science and technology themselves were modified that determined what actually happened. Nations did not have to pursue a high-stakes technological arms race because of the existence of nuclear weapons; the motivation for that lay in realms other than science and technology—particularly in the traditional views of nations toward power and its role. Developments stemming from science and technology provided the means for an arms race, and the nature of the whole system that coupled science and technology with military goals made it all but inevitable that every avenue would be pursued to seek technological advantage and the will-o'-the-wisp of military advantage.

The buildup of nuclear forces also contributed centrally to the emergence of bipolarity, a defining development in international affairs of the postwar period. Only two powers in the world had the economic capability and the technical competence to develop large arsenals of nuclear weapons and their necessary delivery systems, and to maintain the commitment to a high level of continued technical competition. It was not only the military power of what were subsequently dubbed the "superpowers" that led to bipolarity; it also depended on their predominant economic position, the ideological competition between them, and the weakened state of the other combatants of the war who were looking for protection and help.[12] Nevertheless, the dominant military capability of the United States and the USSR, especially in nuclear forces, was a major factor in reducing the prewar system of multiple major powers to two competing centers of power.

That era in international affairs has clearly ended. The economic collapse and dismemberment of the Soviet Union, and the breaking away of the nations of Eastern Europe from Soviet control, signal the emergence of new relationships that are still far from clear. Russia and possibly others of the Soviet Union's successors will retain strategic nuclear power, but the meaning and nature of the competition with the United States, and with other nations of lesser military capability, will certainly be quite different during the 1990s and beyond from what existed between 1945 and 1990.

Some have argued that bipolarity was eroding in any case because of the steady diffusion of conventional military forces and the relative decline of the economic dominance of the USSR and the United States.[13] But whether the argument is valid or not, the collapse, rather than decline, of one of the poles is substantially altering the international configuration of power.

However, the fundamental fact will not change: the most destructive elements in military forces cannot be employed in actual warfare as a means of protecting the state. That does not mean that nuclear weapons will be irrelevant to the unfolding of political developments, for their political meaning as a visible token of power remains. In fact, the disposition of the roughly twenty-five to thirty thousand nuclear weapons of all varieties deployed in the former USSR is a source of deep concern.[14] But the irrelevance of nuclear weapons in actual warfare is a major evolution in the traditional concept of the purpose and use of military force and will remain a consequential element in future organization of the international system.

Scientific and Technological Factors in Strategic Nuclear Relationships

In the light of the continuing rapid changes in international politics following the disintegration of the Warsaw Pact and the Soviet Union, it may seem unnecessary to dwell further on strategic nuclear relationships, a subject that has been so central in postwar international affairs. But that judgment would be premature. The evolving international system may result in a political structure that allows large reductions in the numbers of weapons and changes in their characteristics that will decrease the danger of the resurgence of a nuclear-arms race. To that end, Presidents Bush and Gorbachev announced in September and October of 1991 their intention to make substantial unilateral cuts in nuclear arms.[15] The subsequent dismemberment of the Soviet Union will undoubtedly lead to further reductions. Some, however, have argued that nuclear weapons will inevitably continue to be central elements in the arsenal of many states, perhaps more important than at present.[16]

Whether the nuclear-arms race continues or not, the factors that go into maintaining a stable nuclear relationship, and knowledge about what would undermine that stability, will continue to be relevant for the future. Any new configuration of forces or alignments—possibly involving new nuclear powers—will raise issues associated with nuclear relationships. In any case, it would be naive and imprudent to assume that the world has seen the end of strategic nuclear confrontations.

From that perspective, it would be well to identify the most significant scientific and technological factors that influence strategic nuclear relationships, and that will continue to do so whatever the structure of nuclear forces in the future.

The Spread of Technological Information and the Imposition of Export Controls

One such factor stems from the inevitability of the eventual spread of technological knowledge. Information and ideas underlying new technological capabilities, if not the hardware itself, eventually become known, sometimes quite quickly—through publication, word of mouth, natural exchanges among scientists and engineers, legal or illegal data collection, licensing, reverse engineering, or simply the demonstration that a particular concept is feasible. The growing dependence of military technology on what has been developed for the commercial market, a reversal of the "spin-off" relationship that prevailed until recently, further contributes to the widespread availability of knowledge with military application.[17] Moreover, the rise of technological capacity in a larger number of countries serves to make the technological knowledge base much more nearly common property.

The spread of knowledge is not instantaneous, and it is conceivable that the time advantage a country gained by a particularly important advance in knowledge could be translated into a strategic-force advantage. Such an advantage would be destabilizing to a strategic balance if it provided an incentive for a preemptive strike while the advantage lasted (by reducing the likelihood or cost of a counterstrike), or if it could be used as a power-ful threat in political relationships. It is this objective of gaining an advantage in time that was, during the cold war, the effective purpose of strategic export controls on technology.

Until the end of the threat from the Soviet Union, achieving a usable short-term advantage was not a real possibility for either side, for the strategic balance between the United States and the USSR was entirely too robust. That is, the number and variety of weapons on both sides made incremental change irrelevant to the feasibility of a retaliatory strike or to alteration of the risk of escalation. All that was required to deter attack was the *possibility* of response; certainty was not necessary because even a small risk would entail the danger of unacceptably high costs if the weapons were used. [18]

Historically, there have been periods in which the United States had a strategic advantage (before approximately 1960) and other times in which the Soviet Union was able to convince the world that it had a lead (from Sputnik in 1957 until about the mid-1960s, and to some audiences again in the late 1970s). Neither side, obviously, used those periods for attack, though that presumed advantage does seem to have had political significance. [19] The statesmen of the world are as taken in by the labyrinthine mythology of nuclear calculations as are the analysts who propagate the mythology; as a result, the analysts appear to be correct.

But what of technological surprise? Is it not possible that an unexpected development, arising from science or from synergism among technologies, could result in a very large short-term lead—a "breakout," in current terminology? Surprise is always a possibility with respect to advances in knowledge. But knowledge has to be translated into weapons or other capabilities to make a difference in military power, and that takes time. Not only does that lag allow the knowledge to diffuse, it also allows the consideration of countermeasures, which will be realizable in a similar time frame.

In fact, it is hard even to imagine any advance in technology that could have made more than a marginal difference in the strategic equation that existed between the United States and the USSR up to 1990, unless there was no response at all. Any substantial change in actual capability would require not a single technological advance, but a major systems development (see the discussion of a missile defense later in this chapter), which would add time, complexity, and great operational uncertainties, at the least. It is relevant that the very scope and scale of the r/d committed to military goals by the United States and USSR served to rule out the likelihood of an unexplored avenue available to either one or the other, while protecting against surprise by keeping both aware of technical possibilities. Ironically, as long as this situation continued, r/d reduced rather than increased the probability of a surprise of sufficient magnitude to be destabilizing.

This does not mean that technological change will be irrelevant to the strategic relationships among nuclear powers. Anything like the present commitment to military research and development guarantees that the environment will never be static. Even if an improved political climate leads to a slackening of military r/d, developments in fields as far removed today from weapons application as superconductivity and biotechnology, for example, will surely have military uses in time. That will lead to continued incentives to insure against surprise through r/d, and possibly to changes in tactics, weapons, and deployments. As a minimum, science and technology and their applications will be constantly modifying the knowledge background of military relationships (and modifying security matters generally) and in turn will be feeding the motivations for more r/d.

Another factor that is often ignored is that the acquisition of technical knowledge is not tantamount to being able to use it. Whether developed internally or transferred from abroad, knowledge is successfully applied only if there exists the capability to understand it, adapt it to local conditions and needs, and reproduce it or use it to manufacture products. For a variety of reasons, which will be discussed in the next chapter, the Soviet Union had difficulty in mastering the skills of production of many high-technology items important for military applications.[20]

However, strategic military balances depend not on equivalence in technological sophistication, but on equivalence of capabilities, such as deliverable weapons, survivability, or size of forces. The USSR showed it was adept at offsetting technological shortcomings through other means— for example, deploying larger rockets that permitted the use of higher-yield warheads to compensate for lower accuracy.[21] Superior military technology can make a difference to a strategic balance, but the significance of the difference depends on whether it can be compensated for or countered by other available means, as the Soviet Union was able to do.

Although technological advantage is not the crucial measure of overall strategic relationships, it is an important aspect of the relative capability of operational forces, especially in non-nuclear arms. Recognizing this, the United States and its allies in 1949 placed controls on the export of technology and technological information to the eastern bloc. The controls had the primary purpose of denying military-related technology to the Soviet Union in order to maintain a lead in the capabilities of strategic forces.[22] Concern over the issue greatly intensified in the late 1970s and particularly in the 1980s during the Reagan administration, when there were benighted accusations of a "hemorrhage" of scientific and technological information to the East.[23] At that time the export legislation in the United States was strengthened, and the operating capability of CoCom, the informal mechanism Western nations created for coordinating technology embargoes, was improved. The disintegration of the eastern bloc and the declining threat from the republics of the former Soviet Union led to a start on a substantial easing of the controls in 1990; it was also hoped, in one of history's more stunning ironies, that actually encouraging importation of advanced technology into Eastern Europe and the independent republics would help those countries economically.[24]

This effort at controlling the flow of technology proved to be particularly problematic, and quite possibly a net loss to the United States.[25] Total denial of technology to a country with the capability and will to acquire it is not possible. Even a time advantage in developing or acquiring a technology became harder to achieve as other nations improved their technological capability, as new sources of technology emerged outside the competing blocs, and as military technology came to rely more heavily on commercial advances. Moreover, the Soviet Union, with its command economy, proved to be ill-equipped to make rapid and effective use of sophisticated technology even when it could obtain it on the open market or through other legitimate means.

At least as significant, the efforts to prevent the leakage of technology were themselves costly. They affected industrial trade opportunities—a source of much concern among American businesses, which feared sacrifice of markets to other nations not covered by the controls or not applying them with equivalent commitment.[26] And the controls raised difficult political problems among allies, as differences of views, of legislation, and of goals became focused on easily sensationalized incidents.[27] More important, for it struck at the very purpose of the controls, was the possible cost to the vitality of the scientific and technological enterprises of the West that thrive on the unfettered movement of scientific information, on open industrial competition, and on the free movement of people and ideas. Export controls tend to undermine that vitality, and thus threatened to damage Western superiority in technology more than they denied technology to the Soviet Union.[28]

This is not to say that some controls were not appropriate. It was important to protect operational information and data (cryptographic information, for example), to protect specific technological advances on critical items the Soviets would have had difficulty reproducing on their own, and to impede the flow of information to rogue third-world countries such as Iraq and Libya. But controls have costs, and the costs must be weighed against the benefits that might realistically be achieved—benefits that are often exaggerated because of limited understanding of the characteristics of technology and of the movement of technological information. Aggressive efforts to block the movement of technology permanently, or even for long periods of time, against a determined and technologically sophisticated nation are ultimately futile and may well be counterproductive.

Controls directed against less technologically advanced countries—imposed to discourage arms proliferation, for example—can be both more effective and less costly. Even then, technological information cannot be permanently denied; other means to prevent transfer and use of dangerous technologies are required.

Surveillance Systems

One of the more spectacular outcomes of science and technology and their applications has been the development, from the 1960s onward, of enormously enhanced, often space-based, surveillance capabilities for military (and commercial) use. These are capabilities dependent on technological developments in a wide variety of fields, including rockets, propellants, space, computers, communications, electronics, optics, materials, and many others. Surveillance systems can be said to have revolutionized the gathering of intelligence information, for they make possible real-time observation of an adversary's forces levels, changes in deployment, internal communications, and force movements that simply was unavailable on a regular basis in the past.[29] Those technologies, whose actual capabilities are still obscured by secrecy, have come to play a central role in the collection of intelligence and the verification of international agreements, both discussed later in this chapter. In strategic relationships, they have had special significance for the maintenance of stability.

Their contribution to stability comes primarily from their value in reducing the danger of surprise, either of sudden changes in an opponent's force capabilities or of actual attack. Substantial changes in deployments of weapons or of forces, or preparations to mount military action, can no longer be totally hidden from optical or electronic observation, allowing a potential adversary to take precautions and countermeasures. The likelihood that monitoring is occurring is itself a deterrent to undertaking operations that can be perceived as threatening and offensive.

There is another side to the coin, however, as is typical with any technology. The performance of technology-based systems can never be assumed to be perfect, especially as they become increasingly complex. Human operation adds uncertainty and the further possibility of error.

Countermeasures to degrade or fool a system are possible; so, too, are electronic malfunctions, unexpected interactions, and "common-mode" failures (the correlated failure of components—that is, one failure's leading to another, or concurrent failures due to a common event). Or the output may be misinterpreted by expert analysts, who may draw conclusions that are incorrect but difficult to challenge—especially in a climate in which worst-case analysis thrives. Redundancy of information and systems is a standard way of protecting against such eventualities, as is ensuring that forces are designed to be survivable in a sudden attack. But the possibility of error or malfunction, of misinterpreting a benign signal as an offensive action, cannot be completely ruled out; and such a failure could precipitate the very conflict the surveillance systems are designed to prevent.[30]

There is also the possibility that one country might be tempted to launch a preemptive attack if its surveillance systems showed another country was on the way to acquiring strategic superiority. It is hard to take this argument seriously for major nuclear powers, for the threat of unacceptable retaliation remains unchanged; but it could be relevant in other potential conflict situations, especially third-world confrontations, in which initial differences in capability could be quite large.

Notwithstanding these negative aspects of surveillance systems, there is no doubt that their deployment in the 1960s proved to be a powerful force for maintaining the stability of the strategic relationship between the United States and the Soviet Union. In fact, surveillance systems will be of continuing value in the new situation after the cold war, as the United States and Russia contemplate the establishment of a joint early-warning center.[31] This pooling of data about missile launches could serve as reassurance for those two countries, as well as providing improved ability to track missile launches from third-world countries. Being able to identify immediately the site from which a missile was launched will help to deter the use of such weapons by making swift retaliation more likely.

The integration of surveillance systems with high-technology systems for the command and control of nuclear forces raises some troubling questions, however, as we will see below.

C^3I and Close Coupling

If there is any part of the defense system of nuclear powers that is more dependent on advanced technology than any other, it is the complex of capabilities necessary for the command and control of forces. The shorthand form C^3I is used to denote these capabilities, encompassing command, control, communications, and intelligence (one facet of which is surveillance capability). In many respects, these functions are the Achilles' heel of strategic forces, for the ability to take measured actions in a crisis, to respond to attack, to prevent undesirable action, and to attempt to limit damage in the event of hostilities all depend on the adequacy of information available to government officials and on their ability to impart orders with the confidence that they will be correctly executed.[32]

Some of the vulnerability is unavoidable, for the officials with the responsibility to make the life-and-death decisions of nuclear war are necessarily limited in number, are in more or less public view, and cannot be physically protected at all times. Communication links between them and sources of information or forces to be commanded are at times dependent on exposed antennas, fragile communications equipment, or vulnerable systems in space. Command posts, few in number, are difficult to protect against determined attack and are unlikely to survive for protracted war management.[33]

The United States' vulnerability, according to Bruce Blair, arises also because of the relative neglect of this aspect of the strategic deterrent. In 1985, he argued that

> the overall strategic balance has been relatively unaffected by changes in the size and technical composition of the U.S. and Soviet nuclear arsenals. The state of C^3I has been the *primary* determinant of overall strategic capabilities, and thus the United States (and doubtless the Soviet Union) has lived with enormous vulnerability for a long time. *Exposure to potentially incapacitating attack against command facilities has been the salient feature and central problem of the strategic situation* for nearly two decades. Major developments in the areas of perceived importance—size and technical composition of strategic force deployments—have not had nearly as much effect on the strategic balance as many have believed or alleged.[34] [emphasis added]

Blair goes on to point out that there is much latent instability in such mutual vulnerability and that the fact that "programmed emergency operations of nuclear organizations are geared for launch on warning is symptomatic of this instability. Mutual command vulnerability creates strong incentives to initiate nuclear strikes before the opponent's threat to C^3I could be carried out."[35]

This is a strong and disturbing conclusion, made more so because the subject is so technically complex and so obscured by the secrecy surrounding it. Blair may overstate the incentive to initiate a preemptive nuclear strike on the basis of a perceived threat to C^3I. Even a low probability of nuclear retaliation should have the effect of preventing a rational leader from initiating such an attack, as has been the case without serious challenge since the mid-1950s. Nevertheless, the vulnerability of command and control and its effects on the stability of a nuclear balance is a subject of great significance to strategic relations, and one that has been largely outside the purview of public debate.

The end of the direct confrontation between the United States and the fragmented USSR may make the vulnerability of C³I a less immediate matter, but command and control issues will remain serious as long as nuclear forces exist. Obviously, any resurgence of nuclear confrontations, whichever states are involved, will bring these questions to the fore once again.

There are two particular points to be noted about the influence of science and technology on C³I. The first is the most evident: C³I is peculiarly dependent on technology for the systems that constitute it and for the assessment of both its vulnerabilities and the consequences of those vulnerabilities. Given its highly technical nature and strategic importance, it tends in the United States to be the province of a small number of technical specialists and policy officials; yet it is as important an issue as the structure of strategic forces, a subject debated and redebated long and hard in the press, the government, and the academy. It certainly received even less public airing in the former Soviet Union.

The second point has to do with the increasing use of technology for real-time monitoring of an adversary and with the growing technical complexity of military command systems that often require preprogrammed instructions. The result is to create tightly coupled systems that tie adversaries together in ways that lead to dangerous instabilities. For example, a decision to increase the alert status of forces in a crisis situation, a time-honored means of signaling a warning to a potential enemy, now could itself trigger hostilities. Alert status in a nuclear environment might include measures such as preparation of weapons systems for possible firing; dispersion of aircraft or other forces; delegation of responsibility for weapons to officials who are then unavoidably out of reach (e.g., on submarines possibly cut off from communications while deep under water during a crisis); or granting authority for military commanders to fire weapons if communications are lost with the political leadership. Even if alert status is adopted in great secrecy, it will become immediately evident to the other side by surveillance of visible activity, changes in communications and electronic patterns, and other effects. In fact, if the moves did not become evident, they would have no use as a signal. But such moves would be as appropriate for preparation for a surprise attack as for signaling a capability for retaliation, and an adversary would not know which.

The inherent ambiguity of the actions, the overwhelming damage to the defender if the alert turns out to be preparation for attack and not simply a political signal, the reduced time available for considered human intervention, even the possibility of a mistake or loss of control if authority for the use of weapons has been delegated—all conspire to produce a very dangerous and unstable situation. And, it must be remembered, if a move to alert status has been for signaling purposes, it is presumably because a crisis of some kind has arisen. The possibility of misinterpretation is thus even greater.

As the full weight of this situation became clearer, neither the United States nor the USSR attempted to use a change in nuclear-force readiness for signaling purposes after the Yom Kippur War in 1973, though there were later occasions—such as the Korean Airlines incident and the U.S. raid on Libya— when one or the other might have done so were it not for the realization of the danger.[36] The effects of technological change on the structure and characteristics of C^{3}I have thus locked both sides into a closely coupled system, in which moves by one are immediately observed by the other; the actions are unavoidably ambiguous, and the costs are unacceptably high if incorrect judgments are made. This system is a prescription for great instability (Bruce Blair and John Steinbruner describe it as "a fatal accident waiting to happen"[37]), although the high costs of actually firing nuclear weapons impart a countervailing stability.

The situation may be eased by the change in political/military relationships after the demise of the Soviet Union. However, continuing disarray and uncertainty create their own dangers. During the attempted coup d'état of August 1991, the briefcase containing the nuclear-weapons codes was taken from President Gorbachev.[38] That might not have given the conspirators control over the firing of weapons; but after the breakup of the Soviet Union into a number of independent republics, it is not yet clear who will have possession of the nuclear warheads and the means to fire them, or to prevent their being fired. It could indeed be an "accident waiting to happen."

Continuing technological change is not likely to alter this tightly coupled relationship of opposing nuclear forces; it will become more complex and dangerous if additional nations become significant nuclear powers. Measures to reduce the size and distribution of nuclear forces and to avoid nuclear confrontations are by far the preferred, and perhaps the only, routes for dealing with this set of issues so closely tied to the effects of continuing technological change.

The Changing Significance of Time

A subject to which we will return in several different contexts is the role of time in the political decision processes, and how this role has been altered by technology-induced change. This subject is particularly relevant to strategic-force relationships, as is immediately evident in the preceding discussions of the significance of surveillance technology and C³I. The combination of thirty-minute (or less) delivery time for nuclear-tipped missiles and the availability of information in real time but necessarily in large quantities places inordinate pressure on decision systems. The *quality* of the policy process of governments has always been a factor in international relations; now it becomes central to basic decisions that can mean the difference between survival and annihilation.

Not only will individuals, or an individual, have to make global life-and-death decisions in a matter of minutes after receiving information that indicates possible launch of a nuclear attack, but much of that information will have been selected and sorted by preprogrammed equipment. The raw output of surveillance or early-warning systems, is not useful for a decision that might have to be made in thirty minutes or perhaps in much less time. The information must be reduced, interpreted, evaluated; the uncertainties and ambiguities made evident; alternative actions and their consequences spelled out; and possible orders prepared. That cannot be done within a half hour after the intercepts first become available, except by automatic equipment. And the finer the resolution of the surveillance and early-warning systems, the greater the amount of data reduction that is necessary. But preprogrammed data reduction requires advance selection of criteria and anticipation of possible crisis scenarios, and thus dependence on the values and assumptions of those who build and program the equipment; they are the ones who decide what elements of the data are important and how to interpret inevitable ambiguity in the data, such as whether the signals indicate an accidental launch, or a prelude to an all-out city attack, or even an attack at all.

The pressures of time and the nature of the information generated and analyzed by the technology can preempt the opportunity for deliberation and consideration of options by the individual who must make the final decisions, and can bias the choices that are presented. It should be noted that technology does not offer a way out of this situation. The use of technology to assist the strategic decision process would simply mean more implicit delegation of values to the programs built into the technology. The only way out is through political decisions that change the overall nature of a nuclear relationship.

Other Technologies and Defense against Missiles

Other possible technological developments that could alter elements of strategic forces—improved submarine detection, shorter missile burn-time, greater missile accuracy—would have an influence on nuclear relationships; but, as indicated in the discussion about technological advantage, such developments would not upset a rough balance among nuclear powers. The technological development that *would* have a dramatic effect would be one that overturns the dominance of the offense in strategic forces: a defense against ballistic missiles. Since the basis of deterrent strategy is the certainty of the ability to deliver a massively destructive response to an attack, any development that puts that certainty into serious question could change the strategic equation. Though the United States and the USSR long engaged in r/d to explore the possibilities of a missile defense, both governments accepted as early as the 1960s how unrealistic, and unwise, would be the attempt to deploy a full-scale ballistic-missile defense.[39] The Anti-Ballistic Missile (ABM) Treaty became possible in large measure because of the realization of the futility of an effective defense and because of the desire to avoid an unnecessary race to deploy systems that would be inadequate in a purely defensive role but could nevertheless be highly provocative (by raising the possibility that the effectiveness of retaliation after a first strike would be reduced).[40]

The problem standing in the way of the development of an effective defense is primarily that incoming missiles do not have to be particularly good; all they have to be is numerous. The destructiveness of even a small number of thermonuclear missiles (some would say even one) that reach urban targets would produce damage beyond any calculation of national gain; McGeorge Bundy aptly described it as a "disaster beyond history."[41] A strategic defense has to be either perfect or at least able to threaten such a large proportion of incoming missiles as to reduce fundamentally the damage that would be inflicted by retaliation in response to a first strike. But it has to be able to do this in a way that cannot be offset more easily and cheaply by the opponent; the cost-exchange ratio is today overwhelmingly in favor of the offense, and there is no evident prospect of altering that situation for long into the future.[42]

President Reagan surprised the world in March 1983 when, refusing to accept this view of the prospect of an effective missile defense, he called for the development of a defensive shield that would make nuclear weapons "impotent and obsolete."[43] The Strategic Defense Initiative (SDI) was then established as a major program to translate his vision of a 100 percent effective missile defense into reality. In the face of overwhelming disagreement on the part of the scientific and engineering communities about the feasibility of the 100 percent goal, or even of a less-than-perfect defense, the program went ahead, with the commitment of approximately $3.5 billion per year through the remainder of the Reagan presidency and continuation at a somewhat reduced level (because of congressional resistance) in the Bush administration.[44]

What is wrong with SDI, with the commitment to find a way out of the "balance of terror," when technology has been successfully harnessed for other seemingly impossible objectives? SDI illustrates well some of the misunderstandings about the nature of science and technology and about their characteristics and limitations in serving policy goals. The lessons to be drawn are particularly relevant for our study, for in one form or other (though usually less dramatically) they arise often in many policy areas.

The first lesson, applicable to all technological knowledge as it is turned into application, has to do with the difference between the development of a specific piece of hardware and its incorporation into a complex system. The president appeared to believe that a single technology could constitute a defensive shield, something akin to an umbrella that could be raised over a nation or a city and that would be impervious to missiles striking it. It is not possible to prove that such a technology could never exist, but there is no present concept in science or technology that could be the basis for its development—and hence no way to pursue it, no matter how desirable it might appear to be.

Rather, a missile-defense system would require an exceedingly complex mixture of individual systems based on a wide variety of technologies, many of which not only are not yet developed, but are not known to be physically realizable. All of them would have to be developed, tested, combined, tested as an interactive system, fabricated, deployed, modified in the face of real and theoretical threats, and continuously updated. The need to take a systems approach in the design of military weapons is by now well understood (in the military, at least), but in this case the complexity of the final systems would go far beyond any previous experience. Moreover, the dependence at the outset on entirely new scientific knowledge as well as technological development creates, at best, enormous uncertainties; the possibility of easy and cost-effective countermeasures is predictable; and, finally, the system would never be able to be tested as a system. It must work the first time it is used, and work perfectly. We have learned that perfection cannot be assumed for any technology, no matter how careful and extensive its development.[45]

It could be argued that high performance in the complete system is not required, that the mere possibility that it would work properly would be enough to deter an attack. There is some validity in that argument; the theoretical performance of systems carries weight even if far removed from what operational performance would actually be. Missile forces designed for retaliation are assumed to have certain capabilities even though they, too, are never tested as a full system. There is a fundamental difference, however: the operational performance of a missile system intended for retaliation could be degraded substantially from its design capability and would still be able to cause an unacceptably high level of damage; but even a limited degradation of performance of a defensive shield would effectively negate its value for protection of a nation—and its level of performance would be quite clear in advance.

The second lesson, which has wider applicability, is the lack of appreciation of the dynamic nature of science and technology. The discussion about SDI, at least on the part of senior political figures, assumed that this would be a one-time development. Once it was accomplished, the world would be safe from nuclear weapons; to avoid any concern about asymmetrical advantage, the president made the bizarre suggestion that the United States would "give" the technology to the Soviet Union.[46]

In fact, of course, one of the most important characteristics of science and technology is the steady, not necessarily smooth, advance of knowledge, especially under the stimulus of targeted r/d support. Whatever gains were achieved in the technologies that would be part of an SDI system would be constantly modified as new information, concepts and threats became known. The threat aspect is particularly important, for each component technology is vulnerable to countermeasures based in part on the same knowledge that made the component feasible. That is, the knowledge of how to build something carries with it the knowledge of how it might be countered. A constant measure-countermeasure game is characteristic of development of military technology where there are two (or more) adversaries with a stake in the outcome and with the competence to play the game.

The offer to provide the technology to the USSR betrays a lack of understanding of this underlying dynamism and the vulnerability it creates. To know the technological basis of a system is to know how it can be defeated. Thus, the provision of the technology of a crucial system to a potential adversary provides the knowledge to counter the system. That does not mean that sharing of technological knowledge would always be a foolish idea; there may be some conditions under which it would be appropriate and valuable. But an adversarial climate, without accompanying major reductions in armaments (the situation at the time President Reagan launched SDI), is not one of them.

There is another serious problem with the idea of sharing SDI technology, a problem that arises because missile-defense systems can be used to make a first strike more effective just as well as to defend against a first strike. They can play this offensive role by destroying space-based surveillance systems prior to an attack, and by raising the possibility that the destructiveness of a retaliatory strike could be reduced. The idea that the United States would supply technology that could be used to support a first strike against itself by its primary adversary beggars the imagination.

A third lesson stems from the overvaluation of what science and technology can achieve, a natural human response to the "miracles" that their progress has made possible in this century, especially since 1945. The result is a belief that those fields can, under the right conditions, produce the solution for all problems. The right conditions usually imply substantial funding and political commitments like those for the Manhattan Project or Project Apollo. When this belief is applied to social issues—with the statement "If we could go to the moon, why can't we . . . ," completed with references to urban decay, poverty, third-world development, or crime—the inappropriateness of the dependence on science and technology is obvious. Those are dominantly social issues, even if there are technical aspects and ways that technology can help. For subjects that are more evidently technological, such as SDI, the inappropriateness of the idea is not so obvious.

The mistake is that there is not a technological solution to all problems, even ones that are heavily technological in nature. A "Manhattan Project" approach is of no avail if the underlying scientific knowledge that makes a solution possible does not exist. For the development of the atomic bomb, the Hahn-Strassmann experiment of 1938, which demonstrated the splitting of the uranium atom, gave physicists the knowledge that release of the enormous energy of the atom was possible. When the experiments of von Halban, Joliot, and Kowarski showed that more than one neutron was released when uranium was split by a single neutron, the feasibility of initiating a chain reaction was clear.[47] By the time the Manhattan Project was begun in earnest, it was possible to lay out a program to answer the many detailed, but not fundamental, outstanding questions and to develop the necessary materials. Success was by no means assured, but the possibility of success was clear, and the potential routes to it could be fairly readily identified.

Similarly, Project Apollo was technically feasible at the time it was launched. Whether the moon landing would be worth the expenditure of human and financial resources, whether it could be managed successfully, and whether safety could be reasonably assured were all important questions, but the technical feasibility was not in doubt.[48]

In the case of SDI, not only is success not certain, but the knowledge base is not available for some of the necessary concepts.[49] A continuing program of r/d, much of which was in place before the president's speech, was reasonable given the existing understanding of the technical facts. An all-out, high-cost effort to produce an operational system, however, did not have the required level of prior knowledge. Only a political leader at the level of the presidency could have made such a commitment in the face of what was known, or, more appropriately, not known; the fact that Reagan did so says a great deal about him, and about the quality of the technical advice he was receiving (or listening to).

There is another fundamental difference between SDI and programs such as the Manhattan and Apollo projects. In the latter cases, the "opponent" was the laws of nature, which do not change; once they were dealt with, the battle was in an important sense finished. The moon could not take evasive action to avoid the Apollo landing. In the case of SDI, the opponent has the option of countermeasures, of changing the nature of the opposition. The battle is never finished as long as the opponent stays in the game.

The fourth lesson relevant to our inquiry is the belief that science and technology are, even for high-technology subjects, the sole determinants of the outcome, when, in fact, they never are. SDI, whether feasible or not, was conceived in the context of the stability of the existing strategic relationship between the United States and the USSR. President Reagan asked in his speech, "Wouldn't it be better to save lives than avenge them? Are we not capable of demonstrating our peaceful intentions by applying all our abilities and our ingenuity to achieving a truly lasting stability?"[50] But would SDI contribute to stability, even if it were realizable? There are many reasons to doubt that it would. The transition to a defensive mode would itself be fraught with danger, as the implementation of partial defensive systems by one side could be interpreted as signaling an intention to limit damage from retaliation after mounting a first strike, rather than as a means to reduce the damage of a first strike from the other side. The same problem exists at any stage of the implementation of a defensive system if the other side does not have reasonably symmetrical capability. It was largely that reasoning that led to the ABM agreement of 1972.

There is a further important lesson related to the technology of a presumed SDI system. If there is to be any hope of destroying all or most of the missiles after they have been fired, the SDI system must be capable of attack at all points along their trajectory, including during the period immediately after launch, when they are most vulnerable. That implies that the system must be able to be activated within approximately three minutes of launch, while the missiles are still in the atmosphere. Whatever technological ideas there are for such a capability—ideas that include the launch of mirrors into space to reflect ground-based lasers, atomic explosions in space to generate X-ray laser beams that will be aimed at the missiles, the stationing of large lasers and their power supplies in space, and small, independently targeted devices in space dubbed "Brilliant Pebbles"—the implication is that action will be taken essentially instantaneously in response to radar or other warning of attack.[51]

Such a system, able to be triggered by necessarily ambiguous electronic signals, would not be a contribution to the stability of a superpower relationship. Nor, for that matter, is it likely that the American public or

Congress would be willing to accept a system that had to respond without human intervention with actions that carried more than a casual danger of initiating all-out nuclear war.

There are other controversial elements of the SDI debate, though discussing them extensively here would take us beyond our purpose. For example, some argued that pursuing SDI was a way of forcing the USSR into an economy-draining r/d race, or that it could be used as a bargaining chip in arms-control negotiations. Supporters offered both of those justifications, though there is no reason to think President Reagan had either purpose in mind; and he certainly showed himself unwilling to use the program for bargaining purposes.[52] It is unclear whether the program contributed to the realization by the Soviet leadership that the country was unable to keep up with the West in technology, though it may well have added to the general sense of the inadequacy of Soviet r/d, and the perception of the futility and cost of the arms race.[53]

There are also reasonable arguments for a limited antimissile system that would be, in effect, a terminal-phase ABM system improved from 1960s versions of the technology and intended only to protect retaliatory forces. Such a system could be a positive contribution to the stability of nuclear relationships if arms agreements led to greatly reduced levels of intercontinental missiles.

The possible value of SDI as insurance against accidental, unauthorized, or third-country attacks has been used increasingly as justification for continuing development. The ability of the system to work against all such attacks is highly questionable, primarily because of the wide variety of delivery means available for limited attacks. The system's cost would be enormous, notwithstanding its limited effectiveness. Moreover, systems designed to counter accidental or unauthorized launches, which by definition would come without warning, would have to be able to discriminate between launch of hostile missiles and a growing number of peaceful firings of civilian space vehicles.[54] But the apparent success (hotly disputed by some critics) of the Patriot antimissile system in the Gulf War, even though it has little to do with the feasibility or desirability of SDI, has given new impetus to SDI supporters and to the U.S. administration.[55] However, support for the program fell substantially after President Reagan left office and will probably remain lower than either he or President Bush would want.[56]

We have seen in this discussion some of the ways that the characteristic outcomes of the scientific and technological enterprises affect the nature of strategic nuclear relationships, beyond the overwhelming significance of the nuclear weapons themselves. New technologies—surveillance systems, for example—serve a generally stabilizing role. But the time constraints introduced by such technologies, the complexity of the technologies and the data they produce, the corresponding increased reliance on machine processing of information and analysis of policy options—all tend to limit the options and time available for considered human decision in the use of strategic nuclear weapons, and thus inject (in C^3I systems, for example), potentially dangerous instabilities in strategic relationships.

We have also seen that technological superiority, though it can have operational significance, has only limited strategic importance in a robust strategic balance. This is so because change in deployed weapons technologies necessarily takes place only gradually, and the certainty of the spread of new knowledge allows states with approximately equivalent r/d capabilities to maintain rough technological parity. Moreover, the diffusion of technological information can at best be delayed; it cannot be permanently embargoed. Major jumps in technological capability—breakthroughs—that would suddenly alter a strategic balance are not realistic possibilities. Technology is not a magic talisman that can meet any objective set for it; it is necessarily limited by physical principles, the state of knowledge, the inevitable problems of systems design and integration, and finally the limitations of the human systems of which it must be a part.

These observations apply primarily to strategic relationships between or among major nuclear powers. We turn now to consideration of how scientific and technological outcomes influence the evolution and diffusion of military power more generally.

The Evolution and Diffusion of Military Power and Capability

While the world's attention has been focused on the development and deployment of nuclear weapons in the strategic contest between East and West, a revolution of a more incremental nature has been taking place in conventional weapons, a revolution that may over time also have profound effects on international relations.[57] Whether these effects will alter the role of military force as significantly as the advent of nuclear weapons is at present unclear, but they will certainly affect the consequence and danger of "conventional" conflict.

The Evolution of Conventional Weapons

Since World War II, the largest part of the r/d funds committed to military purposes in the United States has been devoted to improving and expanding the capabilities of conventional forces.[58] As with r/d for strategic weapons, the nature of the interaction between the military and technology placed a premium on the development of systems of high technological sophistication (and glamour) and thus also of high cost and complexity. Striking improvements were achieved in the performance of aircraft, tanks, ships, submarines, and personal arms, and equally striking improvements in the weapons designed to defend against or attack them. Some came from spin-off of developments in nuclear weapons, such as miniaturization, computerization, and improved accuracy.

As a result of the ineffectiveness of some high-technology weapons against determined guerrilla forces that was shown in Vietnam, increased r/d attention was directed to applying technology to the conventional battlefield. The "electronic battlefield" of sensors, antipersonnel weapons, and precision-guided "smart" munitions followed.[59] Many such weapons were prominent in the arsenal of the allied forces in the Gulf War. More generally, commercial technology has in many areas moved ahead of military technology and is increasingly available for application to conventional weapons.[60]

The result has been spectacular developments in conventional-forces capabilities: increased and more effective firepower, much greater accuracy, enormously enhanced surveillance and control, higher mobility, lighter weight, and longer range, among others. In turn, these capabilities make possible, *inter alia*, application of force at a greater distance, stand-off offense (or defense), increased firepower for the foot soldier, greater destructive potential, higher probability of destruction of targets, improved defense against fast-moving aircraft, greater ability to suppress air defenses, and enhanced ability to destroy high-value targets, such as ships.[61]

Not only have the effectiveness and usefulness of weapons been increased but in many systems costs have been reduced in relation to the mission to be achieved. Though funds necessary to procure ships, aircraft, and tanks continue to increase, others weapons, such as certain hand-held munitions and antiship missiles, have much lower cost-benefit ratios than comparable weapons in the past. The ships that an Exocet missile, for example, can destroy cost many times more than the missile, the overall value of the Exocet then depending on the probability of success with each missile.[62] The same is true of the Stinger antiaircraft missile.[63] Moreover, the increased use of commercial technology has reduced the costs of technologies useful for conventional military applications, such as computers, navigation systems, and missiles themselves.

Finally, the increased flexibility that is one of the "natural" outcomes of the scientific and technological enterprises has in turn expanded the flexibility of weapons design. The development of plastic explosive (such as the Czechoslovakian-manufactured Semtex) able to be shaped, molded, and painted for ease of transport and deception has added measurably to the capabilities of guerrillas and of terrorists. The design of binary chemical weapons, involving two benign liquids that are lethal when mixed, was intended to make the weapons safer to manufacture and transport; but it will have the additional effect of making chemical weapons easier to acquire and use, available to a much larger number of nations, and less amenable to control by international agreement and monitoring.

The results of this upgrading of conventional weapons were evident in the Gulf War of 1991, with sudden (often exaggerated) recognition in the world press of the revolution in conventional high-technology weapons.[64] In fact, much of the technology used in that conflict was well behind the more advanced technology even then available but not yet incorporated in weapons.[65] Undoubtedly, the result will be additional pressure for increased r/d on conventional arms and intensified efforts by third-world countries to gain access to improved high-technology armaments.[66]

This high-tech "revolution" can only continue, with steadily increasing capabilities for fighting and destruction. Though the weapons become generally more technologically sophisticated, many also become easier for nations with relatively lower technological capacity to operate (and sometimes produce). Certainly the arms trade will, as discussed later in this chapter, make most of these weapons developments available, shortly after their deployment, to most countries that want them.

What will be the ultimate effects of this widespread availability of high-technology weaponry? Presumably, some of the same issues pertinent to military relationships among the major technological powers will apply. Technological advantage over a potential adversary will be seen as politically and militarily significant. That will fuel technological arms races and the development of countermeasures, while raising the specter of instability; nations, emboldened by presumed superiority or fearing imminent inferiority, may decide on preemptive attack.[67]

Instability may also result from the capability for inflicting greater damage in a single attack and from the increased range of weapons, which can enmesh a larger number of states in a conflict—especially in the volatile Middle East. The ability to keep a conflict from spreading may be lessened, as the range of weapons and their destructiveness bring larger powers into the fray (either accidentally or by design). If the level of damage that can be inflicted in response to an attack becomes high enough and certain enough, however, hostilities may become sufficiently dangerous as to introduce a level of deterrence—and thus stability—closer to that prevailing among nuclear powers.

The gap in military capability between major and minor powers would not be likely to change significantly, except that smaller powers would be able realistically to threaten blackmail attacks on, say, large cities, with conventional weapons of substantial destructive potential. Obviously, possession of nuclear, chemical (CW), or bacteriological (BW) weapons would make such threats more credible and thus of greater concern.

It is well to note, in addition, that technological sophistication in conventional weapons, as in the case of nuclear systems, tends to compress the time available for the exercise of human judgment in the use of the weapons. This introduces a potentially dangerous level of instability in times of confrontation or conflict, as became quite evident in the battle–tank losses to "friendly fire" during the Gulf War.[68]

Could the escalation of conventional-weapons capabilities through technology ever reduce or eliminate the resort to war as an option to advance national objectives, as it has for nuclear weapons? That is a much larger question that has some intellectual appeal for a distant future, but only modest conviction for any foreseeable future. Standoff fighting capabilities and large destructive potential may serve to deter more local wars than in the past, but the political forces that lead to conflict are not likely to be defused for long by the threats posed by conventional arms.

Perhaps the more immediate issue is that more-effective weapons will be found in the hands of rump portions of states.[69] The internecine warfare in Yugoslavia; the contending elements of India, Iraq, and Ireland; and the disintegration of the Soviet Union are clear and disturbing examples. Technological sophistication of weapons could make the effects of such developments very much greater than in the past, both within those nations and for other states.

The Diffusion of Technical Knowledge

One of the essential characteristics of technology—the unavoidable diffusion of technological knowledge—has a direct bearing on the significance of escalating capabilities of conventional weapons.

Sooner or later, the knowledge that can be translated into weapons can be acquired (or developed independently) by any country with the technical competence to understand it and the resources to invest in it. Eventually, the weapons themselves will be able to be built if the choice is made to do so. There will be delays: some processes for a time may be too difficult to master, special materials may have to be located and acquired, political considerations may deter development or production, and many other factors will naturally affect weapons-production decisions. But the necessary knowledge will sooner or later be available even for what were once considered to be highly sophisticated weapons systems. In fact, with continued development over time, most systems become easier and less expensive to produce or copy.

As a result, an increasing number of countries, particularly third-world countries, are producing advanced fighter planes, ballistic missiles, sophisticated equipment for ground troops, and other military systems that were at the forefront of capability when first developed, primarily by the United States and the USSR. In 1991, it was estimated that eight developing countries could build fighter aircraft, six could build tanks, and six could build battle helicopters; by 2000, fifteen would be able to build missiles.[70] The U.S. Office of Technology Assessment reports that in 1988, six present or former developing countries were producing forty-three different major weapons under international licensing agreements; several have entered the arms-export business, helping to create a buyer's market in which weapons are generally available to any nation that can pay for them.[71] Among the more disturbing examples is the diffusion of the capability to produce BW or CW weapons, in plants that can also be claimed to be for peaceful industrial purposes. Those weapons, in the hands of rogue states such as Iraq and Libya, give aggressive and erratic leaders the capability for major destruction and terror.[72]

It is clearly the capability for indigenous production (and then sales) of advanced weapons systems that will make any attempt at control of conventional weapons extremely difficult.

Arms Trade

The most significant factor in the rapid diffusion of military capability is the willingness, even eagerness, of weapons-producing states to sell their weapons to serve foreign-policy goals or improve their trade balance, or to allow their citizens to sell weapons for the large profits possible. Weapons transfers, not exactly a new phenomenon in international affairs, are not inherently good or bad. They can serve to stabilize dangerous confrontations, to advance important strategic objectives, or to support the security of friendly states, just as well as they can increase unwanted tensions and instability.[73]

The sale of arms to Israel certainly has been a factor in the survival of that state, while the provision of Stinger antiaircraft missiles to the Afghan resistance proved to be spectacularly successful in denying control of the air to the Soviet Union.[74] The effectiveness of the Stingers contributed directly to the first defeat of the Soviet Union by a third-world force and to the disillusionment in that country with its Communist leadership.[75] On the other hand, the arming of Iraq with a vast arsenal made aggression against its neighbors a more viable option for Saddam Hussein, and it required the response of the United States and its allies to the invasion of Kuwait to be massive in scale, at great cost in dollars and lives.[76]

Even though there may be salutary motives for some arms sales, the overall risk associated with the arms trade is growing. With the increasing power and cost-effectiveness of conventional weapons and the larger number of disparate producers, the world is being flooded with more dangerous armaments under conditions in which control is becoming steadily more difficult.

At the same time, the economic incentives for sales are larger than ever. The large expenditures for r/d and weapons development lead countries and industry to seek to recoup those costs, while the rapid evolution of conventional weapons, by comparison with earlier eras, creates large numbers of obsolescent weapons to be disposed of and higher development and production costs requiring larger production runs to reduce per-unit costs. Sadly, the surplus of weapons and arms-production capacity as the cold war winds down will add to the number of weapons available for sale; from 1984 to 1988, before the end of the cold war, the United States and the USSR already accounted for 65 percent of all arms transfers.[77]

The Gulf War has also been a stimulus to arms sales, as the value of technological superiority became blindingly evident. It also, however, clearly demonstrated the danger of the uninhibited proliferation of arms, whether conventional, nuclear, chemical, or biological. States have responded to the lessons of the war in contradictory ways: President Bush announced an ambitious weapons-control plan for the Middle East on May 29, 1991, while the secretary of defense said one week later, on June 4, that new arms sales were planned for the region.[78] Export controls on their way to being eased at the end of the cold war are instead being strengthened for some technologies to prevent their falling into the hands of "bad actors."[79]

The primary weapons-exporting nations (the United States, the USSR before its breakup, France, the United Kingdom, and China) began in 1991 a projected series of meetings to consider what policies might eventually be adopted to control the arms trade.[80] It will be a long, slow process, for the incentives remain large for both suppliers and purchasers, the sources disparate, the means of control limited, and the characteristics of technology inimical to effective suppression.

New international initiatives, such as the creation of the Missile Technology Control Regime in 1987, indicate a new appreciation of both the importance of the problem, especially for some categories of weapons, and the necessity of creating new forms of international machinery with a chance of bringing weapons proliferation under control.[81] Undoubtedly, any further progress on controlling arms flow will require genuine determination on the part of the larger states to reach and enforce agreements, along with innovation in the functions and authority assigned to international organizations. (Arms control is discussed later in this chapter.)

There have been two other major international developments with regard to conventional-weapons control that give some sense of optimism. The first is the Conventional Forces in Europe (CFE) agreement, reached in December 1990, which has resulted in large conventional-force reductions of the previously massive presence of the now-defunct Warsaw Pact.[82] Though the agreement was in many respects simply a ratification of the already-changed political and military situation in Europe, it did also reflect the change in climate brought about by the Conference on Security and Cooperation in Europe (CSCE), by 1992 a forty-eight-nation European conference (but including the United States and Canada) first held in Helsinki in 1975.[83] The CSCE has become a valuable forum for East-West discussion and may in time develop a continuing institutional role in European security. The CFE agreement, in any case, and whatever institutional development does follow, is likely to be of immense long-term importance for the stabilization of Europe, the flash point for so much conflict over hundreds of years.

The second international development may in time prove to be of at least equal significance, and the most interesting from an international political perspective. It is the intrusive role that has been given to the UN in monitoring Iraq's weapons deployments and r/d installations *after* the Gulf War.[84] It signifies an unprecedented level of accord among nations to grant meaningful oversight powers to an international organization, powers that have been energetically exercised. That level of agreement and right of unfettered inspection within a nation are likely to be required in the future if widespread control of the proliferation of arms is ever to prove feasible.

Nuclear Proliferation and New Nuclear Powers

The spread of nuclear technology and weapons is a special case of the general issue of weapons proliferation. It is affected by the same factors; however, the dangers of the proliferation of a weapon of such power have led to a much more substantial political commitment on the part of the international community to attempt to prevent its spread.[85] The Non-Proliferation Treaty (NPT), signed by 135 countries by 1990, not only obligates signatories that do not have nuclear weapons to refrain from developing or acquiring them, but also establishes authority for inspections to detect diversions of fissionable material from peaceful to weapons purposes.[86] An international organization, the International Atomic Energy

Agency (IAEA), has as one of its main purposes the performance of this "safeguards" responsibility.[87]

The primary concern over nuclear proliferation is the danger of the use of nuclear weapons in internal, local, or regional crises that could, out of confusion, uncertainty of motives, or willful action, escalate to involve major powers and threaten the entire globe. In fact, the simple acquisition of the weapons for the armed forces of several states could erode the "fire-break" between conventional and nuclear arms that has so far existed, and over time contribute to higher levels of instability accompanying even conventional conflicts. In addition, the possibility that nuclear weapons could be coupled with long-range ballistic delivery systems, increasingly available to those countries that want them, raises to a new level concerns over the spread of potentially devastating conflict. The potential use of the weapon for blackmail by renegade states or nonstate groups is also a natural source of deep disquiet.

So far, the attempt to limit proliferation has been reasonably successful. Even though several of the countries of greatest concern have not joined the NPT (Israel, Pakistan, India, Brazil, and South Africa), only India has set off a nuclear device (and it has set off only one); Israel is assumed to have a stockpile of "near" weapons, though it has not tested.[88] The need for special materials has posed something of a barrier, though the growth of the nuclear-power industry and the involvement of more countries in nuclear research have made those materials more easily available if a nation is determined to acquire them. Political factors militating against nuclear-weapons development have turned out in most countries to dominate the technical ones; the political costs of acquiring nuclear weapons have proven to be a more powerful disincentive than was originally expected.[89] The further spread of nuclear capability, however, must be judged as likely to occur sooner or later; Pakistan and North Korea are apparently well on the way to acquiring weapons, and in 1990 Brazil discovered a clandestine A-bomb development project (which it has canceled) under the control of the military.[90]

The difficulty of preventing the acquisition of nuclear technology by a determined regime is underscored by the clandestine program of Iraq. That nation is a signatory of the NPT, but it mounted a program over many years outside the purview of IAEA inspections. It has succeeded in enriching sizable quantities of bomb-grade material through the use of inefficient but well-understood technologies dating from World War II. The nature of the program went largely undetected until the inspections after the 1991 war.[91] The progress that had been made and the methods used have made evident both the range of possibilities available to a determined state and the difficulty of discovery.[92]

The situation is made more difficult by the problems accompanying the breakup of the Soviet Union. With an estimated twenty-seven thousand nuclear warheads and, more important, significant nuclear-weapons laboratories and thousands of highly trained scientists and engineers that the new independent states cannot support, the likelihood of the sale of information or the migration of talent to other countries is quite high.[93] It is sufficiently high, in fact, that it dictates even more attention to the "demand" side of the proliferation question—discouraging or preventing states from deciding to acquire nuclear weapons, rather than assuming that the movement of nuclear information or scientists can be prevented.

Two other technology-related developments will serve to make proliferation more probable, or at least easier. One is the movement toward the greater presence of plutonium in nonmilitary commerce as a result of the reprocessing of spent fuel from nuclear-power reactors. The purposes of reprocessing are to increase the amount of energy that can be generated from nuclear fuel and to provide feedstock for breeder reactors, which would also extend fissionable fuel supplies.[94] The United States had been attempting to discourage non-nuclear-weapons states from reprocessing, both on economic grounds and to prevent proliferation; but it agreed in 1988 to let Japan reprocess fuel originally supplied by the United States.[95] Aside from giving Japan a supply of plutonium, directly useful for making nuclear weapons, that arrangement will involve shipment of plutonium on the high seas: 150 tons over thirty years, "enough for at least 20,000 bombs."[96]

The other development of concern is the announced intention of some non-nuclear-weapons states, notably Brazil and India, to acquire nuclear-powered submarines, which require highly enriched uranium as fuel. The acquisition of such submarines would give those states access to material suitable for weapons, or a rationale for building enrichment facilities for approved "peaceful" purposes that would otherwise not be permitted under the safeguards regime.[97]

Is it conceivable that one or more states might in time develop nuclear-based military power to rival that of the United States, now the only genuine nuclear-weapons superpower? Concern over nuclear proliferation is not the issue here, but rather the possibility of states with the necessary technological and economic capacity making the political commitment to acquire military power on that scale. Of course, it is not simply the acquisition of nuclear weapons that is involved, but also the ability to acquire the necessary delivery and support systems—a large, expensive, and technologically demanding task.

The only realistic candidates for the role of a new superpower, for a long time to come, would be the European Community, China, and Japan. There is no indication that any of them seeks that role at present.

Western Europe would have the scale of resources, the population base, and the technological capacity to build its military capability. A reunited Germany might be able to do so on its own if the European Community did not survive. However, the experience of the two world wars, the precipitous decline of the threat from the East, the secure democratic base of postwar Germany, and the increasing comfort of economic instead of military prosperity make such a move exceedingly unlikely for long into the future. Almost all the pressures will be in the opposite direction, unless a new threat rises out of the turmoil in the East, and the United States has departed from Europe.

China is the least likely of those mentioned to be able to become a full-fledged superpower within at least the next several decades; but it is the nation with perhaps the greatest pretentions to great-power status, and it has the largest population base and technological potential among developing countries. It is also sufficiently homogeneous and centrally organized as to be able to take difficult and costly political decisions.

Japan is the other nation today with both the technical and the economic resources to build a major military capability, including nuclear weapons, if it made the decision to do so. There is no reason to expect such a move; it is expressly forbidden under the peace treaty with the United States and by the Japanese constitution, and it appears to be most unlikely in the context of present Japanese internal politics. Japan has been able to thrive without a large military, in good part because of the protection afforded by the U.S. nuclear umbrella. Nevertheless, the anomaly of the world's second, or perhaps first, economic power voluntarily eschewing military power may not last indefinitely—especially if trade relations with the United States become more acrimonious, as they are likely to do. This is a new phenomenon of international affairs—economic power without military power. Can it continue? Is military force no longer a necessary component of a country's overall power and influence?[98] It must be added that it is not clear what additional influence a major nuclear capability would provide for Japan. It would, however, almost certainly stimulate China to increase its nuclear capability; similarly, a major Chinese commitment would probably be the one development that might cause the Japanese to do the same.

All this is, one hopes, idle speculation. Ironically, however, the increasingly favorable atmosphere for major nuclear-arms reductions could make more likely the emergence of additional major nuclear powers, perhaps even allowing a larger set of candidates. The problem would arise if those reductions in effect lowered the threshold requirements to become a nuclear power. Presumably, arms reductions of that magnitude would for just that reason require parallel undertakings by third countries not to build or increase their nuclear arsenals. Still, such a world would provide less of a barrier for a nation to "break out" sometime in the future, if one should choose to do so. At the least, regional powers might emerge, large enough to dominate military affairs in their region without directly threatening other regions or global relationships.

In examining the evolution and diffusion of military power and capability, we have seen how the characteristics and outcomes of the scientific and technological systems that societies have put in place lead to continued rapid and dramatic stretching of performance and cost-effectiveness of conventional weapons and to the inescapable diffusion of knowledge about them. These effects have implications for many aspects of international affairs, in particular adding to the destructive possibilities of local war, the danger of the spread of conflict, and the power available to rogue states, portions of states, or terrorists. Differences in technological capacity can play an important role in confrontations between non-nuclear-weapons states, in contrast to competition between major nuclear powers of roughly equal technological ability. As a result of all these factors, instability in political affairs is likely to grow, especially among third-world countries, unless the destructiveness and danger of weapons serve eventually to deter adventurism or the ambitions of nationalism and ethnic rivalries.

The inevitable diffusion of knowledge makes the proliferation of conventional weapons highly probable in the absence of strong international agreements to limit such proliferation; the economic and political incentives for nations to engage in the arms trade serve to acclerate the rate at which the weapons themselves spread. The diffusion of knowledge would make proliferation of nuclear weapons equally probable, were it not that the international community has been giving much greater political emphasis to discouraging states from seeking to acquire a nuclear capability. The greater availability of knowledge and of highly trained personnel in the nuclear field as a result of the breakup of the Soviet Union places a further premium on reducing the incentives for states to "go nuclear," rather than on attempting to interdict the movement of knowledge and people.

It is unlikely that new nuclear superpowers will emerge in the near future, though reducing the number of weapons in the arsenals of the major nuclear powers in effect lowers the threshold for achieving major-nuclear-power status. The international community is likely to see instead the emergence of regional powers armed with higher-performance conventional weapons, weapons that will continue to be increasingly available unless there is a significant increase—and none is at present observable—in the political commitment to limit their spread.

Intelligence

The capability for gaining information about another state's military forces and intentions has gone hand in hand with the development of a country's military capabilities. As forces have increasingly adopted advanced weapons technology and become dependent on associated technological capabilities in communications and transportation, so, too, have they sought to make use of advanced technology for obtaining intelligence information.

The targeted development of technology to serve intelligence objectives has as a result become an important element in defense-related r/d; the developments are usually shrouded in secrecy, but occasionally become widely known through accidents or public incidents. The shooting down of the American U-2 in 1960 by the Soviet Union brought worldwide attention to the capabilities of that high-flying spy plane, which had been developed explicitly for overflight surveillance missions.[99] The ability to intercept and decipher coded military communications during World War II, made public only long after, was of critical importance in the prosecution of that war by the Allies.[100] Since then, technology for the intercept of communications and electronics traffic has become a major objective for strategic and tactical intelligence; it now involves satellite systems, terrestrial listening posts, and extensive equipment for conventional forces, as well as massive computer capacity.[101] The requirement for high-speed data handling and analysis of electronic intercepts was, in fact, a major stimulus of government support for the development of computer technology in the United States.[102] With the recognition of the much-improved capacity for intercept of messages, countermeasures to deny intercept or to prevent decoding have also received high priority. One result has been the involvement of theoretical mathematicians in high-security matters, much to their surprise and chagrin.[103]

The increasing emphasis on technology as a primary means of gaining intelligence has not been without criticism. With all of the information available from the Soviet Union through use of technology, supplemented presumably by clandestine collection of more traditional kinds, the economic and then political collapse of that nation was ultimately a surprise.[104] Nor were the interception of communications and monitoring of the movement of equipment into and within Iraq adequate to detect the scale of Iraq's nuclear program.[105] As always, no matter how useful or necessary the technology, human judgment is finally required.

The unfolding changes in global security issues may indicate a substantial shift of emphasis in the role of intelligence in general and of its technological component in particular. If military capabilities continue to decline as a key determinant of national strength for industrial powers, with concomitant growth in the importance of economic competitiveness, intelligence targets for those countries are likely to focus increasingly on economic rather than military matters. Certainly, the private sector will find knowledge of the plans and progress of industrial competitors to be an increasingly important objective, and governments' interests are likely to develop in the same direction. Since outright on-site espionage is not normally acceptable among friendly countries, impersonal technological systems for acquiring information would seem to be more readily employable. [106] Some of these systems may be innocuous (e.g., extensive monitoring of scientific publications), but some could involve, just as in the security sector, illegal clandestine acquisition of information (e.g., intrusive monitoring of a company's internal communications). [107]

A change in focus toward economic targets, however, would raise a host of new political issues that would be harder to deal with as matters of policy. No longer would the primary attention be only on another nation's governmental operations; the private sector would be directly involved, with individuals, corporations, and universities becoming overt targets for the collection of information. This shift could lead to pressure for harsh domestic measures to discourage foreign acquisition of knowledge, and certainly could lead to uncomfortable negotiations and disputes among nations with otherwise-amicable relations. It may be that the technological intelligence capabilities of governments will not be extensively diverted to the economic sector in this way, but some observers believe that for some countries it is an already-accomplished fact. [108]

Thus, intelligence operations, though a time-honored activity of governments, have now been elevated in sophistication and capability—and certainly in scale—by the application of technology. And they have been an important factor in the reverse direction—stimulating the development of technologies with particular capabilities. Intelligence operations have become a continuing background aspect of the relationships among nations, usually of relevance largely to security issues, but now possibly broadening to include other aspects of those relationships. Security matters will continue to dominate, however, as the diffusion of weapons and knowledge creates new centers of power that require monitoring. Moreover, the likely increase in formal and informal arms-control agreements to limit weapons arsenals and proliferation has significant implications for intelligence-gathering technologies, as we will discuss below.

Arms Control

The international control of armaments is appropriately seen as another aspect of national-security affairs, intended similarly to contribute to the protection of the citizenry and the survival of the state. Most of the considerations that apply to the effects of technological change on military forces naturally apply to arms control as well, starting with the basic challenge to national survival posed by nuclear weapons mounted on intercontinental missiles. It is that development above all that has raised arms control to the political prominence it has held to varying degrees since the late 1950s.[109] More recently it is the recognition of the fundamental inutility, as well as danger, of strategic nuclear forces that has served to energize negotiations to limit the growth of nuclear arsenals, and actually to reduce them in size.

Until the signing of the START treaty in 1991, arms-control measures in the postwar period dealt primarily with preventing certain kinds of weapons or weapons-related activities (e.g., anti-ballistic missiles or atmospheric nuclear tests), rather than with reductions in the numbers of weapons. This focus reflected not only the difficulty of agreeing to reductions in the face of military pressure for expansion of both the size and variety of strategic forces, but also the fact that the nuclear balance was sufficiently robust as to be relatively impervious to all but large changes in the numbers of weapons.

The incentive for absolute reductions, curiously, comes, primarily from factors other than how numerous the weapons are: the psychological effect of the bizarre scale of nuclear-weapons deployment and the costs of supporting them, the opportunity to "lock in" the changed military and political relationship of the superpowers, and the need to legitimize the denial of nuclear weapons to non-nuclear-weapons states. Concern over the danger of accidental firing adds to the incentives to reduce the number of weapons.[110]

Paradoxically, reductions down to very low levels of weapons could, as noted earlier, put the nuclear balance in jeopardy. If forces are reduced too far, sudden small changes in force size could upset the balance; or a few weapons acquired by a third country could become a force comparable to that of the major powers. To avoid this instability, some level of what has been called "minimum deterrence" appears necessary— though there are occasional voices, sometimes in high places, that have proposed doing away with nuclear weapons entirely.[111] The START agreement, signed July 17, 1991, after nine years of negotiation, calls for reductions of approximately 30 percent in strategic nuclear warheads and weapons.[112]

The need to control the level and numbers of conventional weapons has not received as much attention in the postwar years, though the demise of the Warsaw Pact and the growing realization of the danger of the proliferation of high-technology conventional weapons have stimulated new activity. The CFE pact in Europe was a major step forward in reducing conventional forces on that continent (and ratifying the new political situation). The ballistic-missile-control regime, MTCR, may be the start of a substantial effort to control proliferation of the most destabilizing high-technology weapons; the scare over nuclear and CW warheads for the Scud missile during the Gulf War is likely to give additional impetus to that regime.

Science and technology have additional relevance to arms control similar to other aspects of their relationship to military forces. One of particular importance is the role of surveillance technologies discussed earlier. Originally developed for monitoring the offensive activities of potential adversaries, these technologies now have a critical importance for the verification of arms agreements. Though they may not be sufficient on their own, these so-called national means of verification—which include all relevant intelligence capabilities—are major components of any verification regime and are essential to make arms control agreements possible.[113] Confidence that agreements can be monitored and are being adhered to is at the core of the negotiation of an agreement, its ratification, and its viability over time, especially as the deployment of weapons is reduced.

Perhaps the most striking relationship of science and technology to arms control, however, stems from the role of r/d as a source of constant change in weapons systems. Arms agreements designed to deal with current capabilities will always be vulnerable to new technological developments that affect the assumptions on which the agreements were reached. Even agreements that established machinery to stay abreast of future developments, such as the ABM treaty, were challenged as the technical (and political) background changed.[114] This is a common problem in any subject, but it is particularly difficult in the case of arms control because of several characteristics of the development and operation of national-security systems.

The primary cause of difficulty is the massive commitment to r/d, which guarantees a significant rate of change in the knowledge base and the technology that can be applied to military purposes. This difficulty is amplified by the political calculus by which security issues are considered and, especially in the United States, by the nature of the policy process. The dominance of worst-case analysis—the belief that it is essential to prepare for the most damaging possibilities of an adversary's behavior—has meant that the price of ratification of an agreement can be quite substantial. The U.S. Senate, for example, has typically required a commitment to expand r/d in the very subject of the treaty, on the argument that the United States must be prepared in case the other side finds ways to bypass or nullify the treaty provisions.[115] Even without such a commitment, it is a natural process, given the incentives built into the military/technological system, to move r/d in directions related to, but not prohibited by, a treaty. The technology of underground nuclear testing made remarkable strides after atmospheric testing was prohibited by the 1963 test-ban treaty.[116] The increasing military relevance of dual-use technologies, developed for commercial purposes but with significant military applications, adds to this expanding knowledge base, which tends to undermine the technical assumptions under which arms agreements for both nuclear and conventional weapons are reached.

These difficulties need not be disabling for progress in arms control. Rather, they mean that agreements must be designed with self-conscious concern for anticipating the ways new technological knowledge might alter its terms. And they mean that arms control must be seen as a continuous political process, not one that will ever produce a "final" product.

Since r/d tends over time to undermine agreements, why not make an agreement to limit or halt r/d, as a way of slowing the pace of change? Most scientists would be aghast at the idea, yet to the layman it may not seem such a poor idea.

An attempt to limit research by agreement would be undercut by some of the basic characteristics of the scientific enterprise: wide dispersion, synergism among quite different lines of research, and unpredictability of the results of research or of the potential applications of results. Which research programs would be abandoned? There could be no assurance that important weapons developments would not emerge from research aimed at apparently benign objectives or in subjects far removed from weapons interests, or that militarily relevant research would not produce important knowledge useful for nonmilitary needs. In addition, new factors—such as the improving scientific competence of a steadily larger number of countries and the increased relevance of nondefense r/d to military technological application—would further conspire to make a research ban of highly limited effectiveness, at best, and essentially impossible to monitor and enforce.

A research ban may be unwise and infeasible, but slowing rates of change in particular technologies is not. The development of weapons systems requires both commitment of dedicated resources and testing before the systems can be deployed. Both can be controlled by unilateral decisions or by international agreement when monitoring can assure compliance.[117] It is not completely unprecedented; the 1963 treaty banning nuclear tests in the atmosphere effectively suppressed the development of the technology for peaceful uses of nuclear explosions, not by banning research but by making full testing of the technology hard to justify (the economic advantages would have had to be spectacular) and subject to important political disincentives.[118] The 1972 ABM treaty also retarded the development of some technologies for years, though the restrictions came under challenge as newer technological ideas emerged in connection with the SDI program.[119] Opposition to such limitations on testing from affected laboratories and industry is certain, but interest in measures to slow the pace of change in selected weapons areas may well grow as the improving political climate reduces the weight of many of the opposing arguments.[120]

In sum, science and technology are and will remain important background variables in arms-control negotiations and agreements, as elements that keep the subject in constant movement, as a focus of attention on their own, and as essential ingredients of the monitoring and verification functions that are crucial to viable agreements. At the same time, the effects of technological change serve as complicating elements that increase the difficulty and complexity of reaching agreements and of maintaining their validity over time.[121]

Some Summary Comments

Evolution in the national-security aspects of international affairs has been quite dramatic under the influence of advances in science and technology, though the resulting changes in the structure of the international system have not been as overwhelming as was confidently expected at the dawn of the nuclear age. New developments in the capabilities and dispersion of conventional weapons presage further evolution; but there is little reason to believe that they will substantially change the picture, except to enhance the role of international organizations or other international mechanisms in the attempt to control weapons diffusion. The primary dimensions of the evolution that has taken place can be summarized in a few categories.

Role of Military Power

Compared to the attitudes and assumptions of many in the immediate postwar era, with the horrors of Hiroshima and Nagasaki still fresh, the deep effects of atomic weapons on the structure of the international system have not been nearly as profound as was expected. Sophisticated statesmen and experienced officials in the United States, such as Dean Acheson and David Lillienthal, believed that a new era was not only possible, but indispensable; it would be an era in which the level of international cooperation and commitment to collective security would alter the traditional behavior of nation-states.[122] It is impossible to read of the work of the committee charged with development of the U.S. proposal for control of atomic energy without being moved by their profound sense of the beginning of a new world order, which they were in the process of designing. Their concentrated efforts to understand the underlying science and technology, with the world's leading scientists as tutors, and then to formulate proposals that would bring the atom under international control, were infused with a powerful atmosphere of optimism and idealism, along with a sense of catastrophe if their efforts failed.[123]

The fundamental restructuring did not happen, nor did the disaster that it was believed would follow failure. Instead, familiar political values came to dominate once again as relations between the United States and the Soviet Union turned sour and the international control agency that was the U.S. proposal at the UN was not accepted by the Soviets.[124] Two hostile blocs subsequently formed, each with growing stocks of nuclear weapons and increasingly sophisticated delivery systems at their disposal. But rather than exploding in unprecedented warfare, nations became increasingly familiar with a nuclear-armed world in which the nuclear weapons came to exist for the primary purpose of preventing their being used. In effect, they could serve the traditional purpose of ensuring the survival of the state only if they were never fired in anger. The upper end of the spectrum of military power became a powerful stabilizing force, but it was effectively rendered impotent for actual use in warfare.

This rather substantial change in the purpose and limitations of military forces was not accompanied by restructuring of the state-centered international system, nor did it mean that military power was no longer relevant to international relations; quite the contrary. Even though nuclear forces could not be used in anger, their relative strength had, or (in what amounts to a self-fulfilling prophecy) was believed to have, significance in the political relationships of the leading nuclear powers between themselves and with other countries. Moreover, military action using conventional forces, between either of the nuclear powers and third states or between non-nuclear-weapons states, remained a feasible policy option in more or less traditional ways, with nuclear weapons largely irrelevant to those confrontations.

Level and Distribution of Power

Even though the role of military power in non-nuclear confrontations retains familiar political dimensions, the consequences of technological change and the diffusion of military technology are altering in important ways the capabilities, and thus the implications, of conventional forces. Two broad characteristics of the changes dominate: military capacity generally available, even to small states, is expanding in attributes such as destructiveness, speed of response, mobility, reach, accuracy, and cost-effectiveness; and the pace of technological change in conventional weapons is rapid, and likely to continue to be so.

These characteristics will have many effects. They will make limited wars potentially more damaging and more nearly like major wars to the combatants, because the increased firepower available and the greater radius of application are more likely to involve rear areas and centers of population. The dangers of escalation of a conflict increase as combatants acquire new long-range weapons and are willing to use them against by-standers that would prefer to remain aloof.[125] Instability can result from the capacity to amass substantial power and delivery capabilities quite quickly, though the possibility of high levels of retaliatory damage could also serve to deter attack.

Rapid technological change will likely stimulate local arms races seeking technological superiority or seeking to nullify superiority, and it will create instabilities as relative military capabilities wax and wane. Technological advances in reducing the size, cost, and difficulty of handling of high-value weapons will make it easier for those weapons to be acquired and used and will give nonstate actors access to explosives and weapons with greater capability and effectiveness. The ready availability of weapons to portions or remnants of fractured states poses particularly dangerous possibilities in the volatile situations in parts of Eastern Europe, the former Soviet Union, and other unstable areas. And, of course, there remains the possibility of eventual greater proliferation of nuclear, CW, and BW weapons.

As important as these developments may be, individually they cannot, except for nuclear proliferation, be seen as changing the meaning and role of military force; they simply continue a process (albeit at a faster pace than in the past) of change in the characteristics of weapons and warfare. Taken together, however, they may well, over time, create dangers of sufficient significance to small and large states as to lead to acceptance of collective measures for control of the diffusion and use of force. If that were to happen, it would signal a substantial evolution in the international system. There are small signs in that direction, but the trend is embryonic at best.[126]

Regional Centers of Power

These changes in conventional weapons will likely have another effect, however: the development of regional power centers, that is, regional groupings around states that amass predominant military power in their area. Over time, a system of multiple centers of power may develop whose dynamics more nearly resemble those of the prenuclear era, but in which the centers are geographically much more widely dispersed. The greater dangers of instability in such a development would be disturbing, especially if nuclear weapons were added to the mix.

Heightened Dependencies

Another significant effect of technology-related changes in the security sector is a much-heightened dependence of all nations, nuclear or not, on the decisions and actions taken by others.

The dependence arose in its most dramatic form in the nuclear balance between the superpowers, in which nations were held hostage, their very survival in the hands of another, without any means to prevent destruction if an attack were made. Dependence is also evident at lower levels of non-nuclear military force, as a result of the greater power and reach of modern weapons systems increasingly available to almost any country.

The realization of this changed situation receives its most concrete expression in the restraint that was shown by the superpowers in the use of force between them, even in confrontations with much at stake, and in the general pressure for formal agreements to codify behavior in order to make it predictable and verifiable. There is nothing new in the fact of dependency among states; the threat of total and essentially immediate destruction entirely at the decision of another is, however, new to international affairs.

The high-technology element in conventional forces will also create new levels of technological dependence. Differential access to the latest weapons capabilities may, in regional or local confrontations, have a direct effect on the relative capacity of forces and thus the relative military power of states. Even major powers will find themselves increasingly dependent on technological developments and hardware from other countries, much of it coming from commercial, not military, sources. The technological capability the United States demonstrated in the Gulf War relied heavily on foreign-made components, raising obvious concerns about vulnerability to disruption if sources chose to interrupt supplies.[127]

Geography as a Geopolitical Factor

Geography has long been considered a major attribute in the assessment of a nation's geopolitical strength. It has many dimensions: size, climate, position, topography, agricultural potential, resource endowment, availability of ports, and other physical characteristics. These are relevant not in themselves but for what they imply about vulnerability to attack, ability to exert power at a distance, availability of indigenous resources to support hostilities, adequacy of communications, and similar capabilities that determine a nation's ability to exert power and influence in the international arena.

The most evident change in the meaning of geography is the dramatically reduced significance of physical location as a determinant of the vulnerability of a nation. The splendid isolation of the United States no longer exists, scuttled by the advent of long-range aircraft, missiles, and submarines. More generally, urban and industrial targets of any nation, even though far removed from the battlefield, are now vulnerable to attack from weapons systems that allow the rapid application of force at a great distance. Not only has the protection of distance been drastically reduced, but so, too, has the ability of a state to remain aloof from developments far afield if its interests are affected.

However, geographical position has by no means been removed entirely as a factor in security relations. The United States, for example, is less vulnerable to the short- and medium-range ballistic missiles increasingly available to third-world countries than are the countries of Europe. Nor has transportation technology, to take another example, made the cost of moving large numbers of troops and equipment independent of distance, or made possible their instant availability from distant locations in substantial numbers. The United States was able to mount a major conflict in Southeast Asia, but at tremendous cost both financially and politically. U.S. troops could be moved to the Persian Gulf in 1990 more easily than in the past, but distance and logistic supply routes were by no means irrelevant to cost and effectiveness. In fact, it took five months to amass the forces necessary to attack Iraq.

Spheres of influence are still largely geographically based, and, as noted, the erosion of the military dominance of the superpowers is likely to allow, or encourage, more regional groupings of states for military, economic, or political motives. A nation's strategic policy, moreover, will be much affected by whether it is near unstable countries, or a continent away.

Human versus Machine Control

One of the more disturbing effects of technological change in the security area is the diminished ability to maintain human control over central decisions on the use of military force. This effect is particularly evident in control systems in which the applications of technology have made human intervention difficult, sometimes uncertain, and necessarily dependent on preprogrammed response. The causes of this development derive directly from the steady march of technology: the reduced time for delivery of weapons of mass destruction; the vulnerability of key command and control facilities and individuals; the increased flood of information that must be processed, selected, and interpreted in severely limited time; and the sheer complexity of military systems and their interconnections. A move to anything like SDI, the defensive missile system President Reagan envisioned, would further exacerbate this dependence on machines.

This potentially destabilizing new element in international affairs has, if anything, made governments more, rather than less, cautious about the dangers of inadvertent triggering of uncontrolled escalation. It has made nuclear forces less useful, even as a political threat, than might otherwise be the case. The withdrawal from nuclear confrontation that is accompanying the changed East-West political climate will greatly reduce this particular danger. But any resurgence of nuclear challenges will make it a serious issue once again; the technological character of strategic systems makes that unavoidable.

Science and Technology

Last are the various roles of the scientific and technological enterprises: critical elements in a nation's security posture, natural contributors to the spread of knowledge and armaments, but also limited servants in the search for perfect security. We have seen how much military forces are now dependent on technology, the ways that technological change affects the use and relative capability of military forces, and the dangers for stability of military and arms-control relationships that the continuing development and spread of technology can bring. We have also seen that science and technology cannot meet all objectives and, correspondingly, how important is the understanding of their functioning, their limitations, and their interaction with political, social, and other forces of society.

What is certainly clear is that science and technology have become central elements in the evolution of the security dimension of international affairs, and are so intimately tied to military power, as well as to economic strength, that their health and productivity have become important determinants of national power.

Four

Economies and Polities

ISSUES OUTSIDE the security sector that are affected by science and technology present a more difficult task for this study.[1] Our daily lives are more directly affected by these technology-related changes than by change in security matters; the negative consequences of change are becoming more visible and controversial; and rhetoric—whether critical or favorable—about the effects of technology increasingly bombards the public, the academy, and governments. The gradual shift of attention in world affairs from military competition to economic competition gives greater political saliency to nonsecurity issues; at the same time, the effects stemming from technological change in those issues are recognized to be pervasive and, at best, difficult to untangle from those relating to nontechnical factors. Even when these interactions are no more complex than those in security affairs (although they almost always are), their presence across a broad spectrum of day-to-day human interests makes them seem so.

To probe this large, rather inchoate subject, it is essential to break it down into smaller pieces that are more amenable to analysis. I have chosen five categories, unavoidably overlapping, that best serve the study's objectives: global integration; economic growth, trade, and competition; North-South transfer of technology and dependency; old and new dimensions; and large systems. We will ask, as with the security sector, how scientific and technological change has influenced the evolution of relevant aspects of international affairs, whether there are particular characteristics of technology that favor certain patterns of evolution over others, and whether any fundamental concepts or assumptions underlying international affairs can be said to have been materially altered.

Global Integration

One of the more dramatic international developments of this century, and particularly of the period after World War II, has been the intensified integration of national economies and societies, which has moved nations to new levels of interdependence. The evidence is all around us in daily life and is a constant rhetorical refrain. Though integration is a continua-

94

tion of past trends, it is accelerating with the advances of science and technology to become a major element of the international scene, one that has come to represent a central characteristic distinguishing the present from the past.

This integration is a product of many and varied forces, with technology most often the essential enabling factor. Advances in transportation and communications, in particular, have stimulated the growth of multi-national corporations, of international trade, and of integrated information networks that make possible instant communications and transactions throughout the world. These and other technologies have contributed to the usefulness and profitability of large technological systems, many requiring international or even global deployment. At the same time, expanding population and wealth and more-intensive use of technology have produced escalating externalities that have had costly impacts across borders, and often planetwide effects. The innumerable interactions among nations that result from this growing integration are by no means confined to those managed by governments; transnational activities of a large and growing volume are conducted by individuals, groups, and corporations, often outside the direct control or even the knowledge of the state.

Undoubtedly, the most spectacular technological source of recent societal change leading to global integration is to be found in the explosive growth of information technologies: the convergence of communications and computer technologies, both advancing at unprecedented compound rates.[2] The phenomenon can appropriately be designated the "information revolution," to draw the parallel with the industrial revolution in its expected significance for human affairs.[3] There have been widespread changes in economic, social, and political structure for which developments in information technologies have already played an important role, with the political unraveling of Eastern European and Soviet governments in the second half of the 1980s (which we will discuss below) providing perhaps the most consequential illustration. In the economic sector, the changes may not have been quite as discontinuous, but they are just as momentous. To cite but one example, service industries in the United States, which are in large part based on the capabilities afforded by information technologies, had by the late 1980s grown to account for 68 percent of U.S. GNP and over 70 percent of U.S. employment.[4] Some argue that information has become "the key to modern economic activity—a basic resource as important today as capital, land and labor have been in the past."[5]

It is not easy to resist such superlatives when considering both the pace of change in information technologies and their actual and potential effects as they are introduced. However, these effects, even within nations, are often qualitative and hard to measure. A 1988 study for the U.S. Congress by the Office of Technology Assessment (OTA) concerned with the potential role of technology in major structural change in the American economy concluded that during the next two decades, at least, only information technologies among other technological possibilities "have the potential to change the performance of the economic system itself." The next sentence, however, noted, "If a revolution of some sort is underway, measuring its impact with any precision has proven to be exasperating."[6]

If it is exasperating on the domestic scene, it is at least as much so with regard to international affairs. There is little difficulty in listing the qualitative changes that have already taken place or will take place, such as the effects of the movement of ideas, knowledge, and news, on the ability of governments to insulate their citizens from international developments. Nor is it difficult to list quantitative facts on transborder data flows, increases in trade in information products, or the scale of international capital transactions.

But it is not easy to go behind the statistics or the general observations to analyze the effects with any precision; all too often, attempts to do so become not much more than breathless clichés. Moreover, it is too easy to lose sight of underlying factors that may not be changing in any substantial way, even if surface relationships are quite different from those of the past. On the other hand, it may well be that the effects of information technologies' interactions with other factors in international affairs are so overwhelming in their significance that they not only lead to evolution in the central elements of international affairs, but may lead (or have led, as some would argue) to more-fundamental discontinuities.[7]

It is evident that this increasing integration of national economies and societies, and particularly the aptly named information revolution, is a development with substantial significance for international politics and for the relations among nations; in fact, extensive theoretical and policy literatures in international relations are devoted to studying the political consequences.[8]

Our task is to go deeper than these literatures typically do to try to understand just how the technological change that has been instrumental in this growing integration of economies and societies actually interacts with and leads to evolution in international relationships. One way to do this would be to assemble a long list of technologies that have been major stimulants of integration and then analyze their particular effects on international politics. That would be an encyclopedic task and would prove to be needlessly repetitious.

It will be more productive for our purposes to examine in detail a small number of specific political and economic developments on the international scene that are significant in their own right and that are fertile examples for analyzing what the actual role of technological change has been. That approach will also be useful in exploring whether there are consistent directions of change that flow from the characteristics of new technology, at least in the examples we choose, and in providing some sense of how the dynamics of technological change are likely to affect the issues in the future. We will use the dramatic changes in the Soviet Union and Eastern Europe as a particularly informative political example, and the rise of global financial markets and multinational corporations as significant economic examples.

The Soviet Union, Eastern Europe, and the Flow of Information

It would be oversimple to argue that the sudden collapse of the Communist governments of Eastern Europe and the demise of the Soviet Union were caused by the steadily easier access in those countries to uncontrolled information from abroad. Yet there is little doubt that it played a substantial role.[9] For years, the East Germans could, and did, watch West German television in their homes; all the nations of the Warsaw Pact were covered by radio broadcasts from Western information agencies; audio and video cassettes moved across borders and were impossible to interdict totally; photocopy machines, telephones, computer linkages, and, more recently, facsimile transmission to and from the West were increasingly available in all of Eastern Europe.[10] Information technologies are not entirely new; the telegraph, telephone, and radio have long been powerful initiators of change. What is new is the larger variety of technologies, their more intensive interconnections, the lower cost of their use, the difficulty or impossibility of complete interdiction, the greater impact of the visual image, and the ubiquity of the technologies in the societies as their costs decrease and their entertainment and economic value increases.

The inability of governments to control the channels of information had several political effects in the East. First, it made it evident that the Communist party leadership was unable to match the West's improvement in living standards, and that claims to the contrary were misleading at best and outright lies at worst. Second, the obvious inaccuracy of information provided by Communist governments to their publics led to widespread cynicism and distrust.[11] Third, the lack of official control enabled a substantial level of communications among individuals and groups within and between those societies, communications outside governmental channels or knowledge. This meant not only that the diffusion of information could not be prevented, but that groups could coalesce, ideas could have a chance to germinate and grow, and alternative values could be propagated throughout the society. Fourth, it allowed knowledge of the dramatic events in each country to be rapidly spread throughout the others, creating a tide of change particularly evident in the astonishing last six months of 1989 as Eastern Europe unraveled. And, fifth, unfettered communication provided a real window on the revolutionary events within each nation that helped to accelerate and make irreversible each revolution as it unfolded. The televised scenes of the celebrations on the Berlin Wall and of the demonstrations in Wenceslas Square in Prague, and the crucial role of the central television station in Bucharest in convincing the prostrate nation of Romania of the reality of the revolution, were all major contributors to the continuing events.[12] Two years later, in August 1991, the inability to close down all unofficial sources of information during the attempted coup d'état in the Soviet Union and the broadcast of a defiant Boris Yeltsin atop a tank made it certain that the coup would fail sooner or later.

Of course, the underlying reasons for the monumental changes in the East go much deeper. They relate to more-fundamental national and ethnic forces, many of which had been forcefully repressed for forty years, some since the 1917 revolution. And they relate to major problems of economic performance, particularly with regard to the inability of a command economy to be competitive in technological innovation (a difficulty discussed later in this chapter). Gorbachev's accession to power in the Soviet Union in 1985 began the process of public admission of the failures of Communist society, especially in economic performance. His commitment to glasnost—the encouragement of candor and open discussion, adopted as a necessary part of the restructuring of Soviet society—also had the effect of allowing those more-fundamental issues to rise to the surface. Once it became apparent that Gorbachev would not use force to maintain the Soviet empire in Eastern Europe, the end of the fear of intervention released the satellite nations from their bondage and allowed the protests, different in form in each country, to reach at least initial fulfillment by the end of 1989. Within the Soviet Union, glasnost contributed to a similar mobilization around old tensions and grievances that could now be openly expressed, leading to unrest and separatist pressures—throughout the Soviet republics—that reached their height after the abortive coup and the subsequent breakup of the union.[13]

Information technologies thus proved to be the handmaidens of the political changes in Eastern Europe. In another 1989 revolution, however, in China, they played an equally important role but with the opposite outcome, at least for the time being. The loosening of the bonds that preceded the demonstrations of the spring of 1989, a step taken largely because of the need to make the economy more productive, led to a degree of open discussion and ferment not seen in China at least since the Communist victory in 1949. The student protests and demonstrations, which called for political reform and for dialogue with the aging leadership, became in part a contest for control of communications and information.[14] The massacre of June 4 in Tiananmen Square was a reassertion of power through the use of naked force, with each side attempting to dominate communications in order to have its version of events presented to the Chinese people and to the world. The noted scholar of Chinese politics, Robert Scalapino, described the struggle in vivid terms:

> Most fascinating was the battle over communications. Never in history was the political significance of the information revolution so clearly revealed. The government had to contend simultaneously with a rebellion of its own media personnel and the determination of the students to provide their version of events. The students and their supporters employed telephones, fax machines, tape recorders, foreign broadcasts, wall posters.[15]

The peaceful revolution in this case lost out to the traditional assertion of power. But it seems clear that the increased flow of communication internally and abroad contributed directly to the student demonstrations.[16] The intense struggle to make sure the world, and China itself, knew what happened on June 4 indicates how important both sides believed their version of that event is likely to be for future political evolution in China, especially after the aging leadership of the time has left the scene.

Thus, we see in these events in Europe and Asia dramatic evidence to support the oft-repeated idea that the information revolution, made possible by scientific and technological advance, has permanently altered certain political relationships, constraining the autonomy of governments to determine policies as they alone see fit and the ability of authoritarian governments to maintain centralized power indefinitely.[17]

The general outcomes of the scientific and technological enterprises have resulted in increased ease and lower cost of access to information; more-effective and broader means for diffusion; new forms of presentation of information with more-dramatic impact, such as television; and much easier availability of the technologies themselves. Many technologies played an important synergistic role—for example, transportation technologies that made possible the rapid movement of people for demonstrations, negotiation, or intervention.

It is also clear, however, that the effects are not only in one direction. The Chinese government was able to maintain control and, through time-honored propaganda techniques aided by information technology, to reassert its authority and convince rural China, at least, of the legitimacy of its arguments for suppressing the demonstrations.[18] In an era not far in the past, the Nazis were able to use the information technologies then available as a critical means of securing and then consolidating control of the German population.[19]

It is evident that the effects of the application of information technology, as with all technology, will not always follow one pattern; the results can be quite contradictory, depending on which technologies are allowed to be introduced, on how the technologies are used, and on their interaction with other societal factors. Is it possible, however, in the case of information technologies, that the technology is not completely neutral? Could there be a general bias in its political effects, a tendency for certain consequences to dominate their opposites—for example, might the technology contribute more often toward decentralization of state power than to centralization, or toward a more open rather than a closed society?

That is a complex subject that has long intrigued scholars and novelists.[20] But the role of information technologies in the events of 1989 in Communist countries gives substance to the possibility that there is such a bias—especially in view of the long and strenuous efforts of the governments of those nations both to restrict the access of their population to information and contacts from abroad and to manage the internal flow of information for purposes of control. The developments in the Soviet Union deserve our closer attention from this perspective, with the work of S. Frederick Starr particularly relevant.[21]

Starr points out that domestic horizontal communication networks, a characteristic of a democratic society, were systematically suppressed in the Soviet Union, from Lenin's reign on, in favor of the vertical top-down communication patterns that tend to dominate in centralized, authoritarian societies.[22] This situation began to change after Stalin's death, as the new information technologies of television and computers and, more recently, video-cassette recorders, fax, and electronic mail were slowly allowed to develop. These technologies flourish best when communications and interaction are unhindered; after Gorbachev came to power, Starr writes, "it is reasonable to conclude . . . that a kind of communications revolution is under way in the USSR, [and that it] is modifying the received communication culture by stressing horizontality and interaction across levels where top-down verticality once reigned unchallenged."[23] Starr accepts the fact that the new technologies were also strengthening vertical communication, but he believes the horizontal effects necessarily dominated the social impacts and could have been suppressed only at a very high price.[24]

This shift in communication patterns in the Soviet Union had five major consequences in Starr's analysis:

1. Information was becoming privatized, in the sense that much more information was circulating in the society without any government involvement; this created more diverse sources of political inputs and thus stimulated the growth of pluralism.

2. Information was becoming increasingly internationalized; one result was that the government had to respond more often to information it could no longer control.

3. Individuation—the importance of the individual—was being enhanced, which Starr believes was one of the must important developments in Soviet society.

4. New groups and autonomous organizations (independent of the government and the Communist party) were fostered by the networking potential of new communications technology.

5. Public values were increasingly being shaped by films, ideas, and other sources not under the control of the Communist party or the government.

Starr, writing well before the events of 1991, concluded that these consequences "may eventually lead to a very different type of political order than has heretofore existed. . . . society may remain partially controlled, but it in turn exercises a control of its own." He believes that "such circumstances impose absolute boundaries on absolute power . . . [and on] the government's ability to shape society."[25] The evolution, moreover, is reversible only at an extremely high political and social cost.

Given the pace of change in Soviet society, Starr's predictions, originally formulated when the changes under Gorbachev existed in embryo only, have proven remarkably prescient. Even then (in 1988), he argued in perhaps his most significant point, that the Soviet Union exhibited many characteristics of a "civil society"—a concept of Western political thought in which government is seen as only one of many institutions in a pluralistic social fabric.[26] What he calls "technotronic *glasnost*" did not create this situation, but it made it possible. (He did not specifically consider that the strains of this pluralism in a nation such as the Soviet Union would be greater than its fabric could accommodate, leading to a fragmentation of the union itself; but it was an obvious possibility from his analysis.)

His observations were made even without consideration of another important aspect of information technologies: their effects when they are introduced for purposes of improving industrial performance—for example, in production control and management. In such cases, the operational modes required to make effective use of computers, for example, also contribute to the strengthening of the horizontal networking that is so important an aspect of pluralism.

Starr's analysis thus provides an explanation for the view that the introduction of information technologies has been critical to the profound set of developments in the former Soviet Union, developments that have had such a large impact on international affairs. His arguments would substantiate the idea that there is a dominant trend in the effects of the introduction of those technologies that favors openness of society and limits on the exercise of authoritarian power. Not all analysts are as certain of the effects as he is, though there tends to be general support in the literature for those conclusions.[27]

Starr does not deal with the other side of the coin—the possible effects on the shaping of public attitudes by combining the new technologies with new techniques of propaganda and advertising, which were so important to the political success of national socialism and the earlier Communist years. New social techniques, coupled with the development and introduction of information technologies that strengthen vertical and inhibit horizontal communication patterns, might make possible new forms of control over publics, or new ways to influence publics across borders.[28] It is a 1984 image, but it cannot be rejected as totally without substance. And, as noted, it is evident that China, a nation with a long tradition of centralized elite government and with a much more homogeneous population than the Soviet Union, has so far been able to use information technologies to assist in the maintenance of control and has been able to resist the development—or been able to contain the social consequences of the development—of horizontal communication networks. As of 1992, power remained centralized in that society as it always had been; and the gradual diffusion of power that followed the attempts to move from a pure command economy had been halted, at least for the time being.

However, the nature of the newer information technologies, the pluralism of information inputs that results, the natural growth of networks outside government control, and the individuation that is a natural product of the new technologies would seem to make it very much less likely that renewed attempts at control of publics through control of information in Europe could have the success that was possible in earlier decades of this century. The rapidity and totality of the change in Eastern Europe and the Soviet Union would also seem to support that view. It remains to be seen whether China will be able to take full part in a world economy and still retain the measure of control of information the present leadership believes essential. The evidence of the student demonstrations in 1989 and of the events in Eastern Europe would appear to indicate it is only a matter of time before the pressures for change become irresistible.

It is therefore a reasonable, though qualitative, conclusion that the introduction of information technologies (and other technologies that play a synergistic role) tends, on balance, to have consequences that are biased in the direction of increased limitations on the centralization of political power and toward greater openness within a society. It is also evident that these technologies can play a crucial role in the propagation of political change in times of rapidly moving events—for example, during the dominolike fall of the Communist governments of Eastern Europe. The effects are not applicable only to authoritarian governments. The effects are easier to see and analyze in those nations, for authoritarian governments, aware of the importance of information to their maintenance of power, have made more conscious attempts to control information flow. But, for all governments, information technologies have similar implications for autonomy, openness, and decentralization of power.

Thus, the effects of these technologies within nations have been substantial, and they have been weighted in one direction. Political commentators tend to accept this conclusion without much question, so that analyses of international affairs typically refer to information technology as a key factor leading to change in the international system.[29] Stanley Hoffmann summarizes the effects as a worldwide trend toward "people power" brought about by the information revolution, which, "on balance, make[s] the control of people's minds and moves by governments more difficult." He argues that it helps to explain the shift in stakes in state preferences and in international goals, as publics participate more extensively in determining the objectives and policies of their nations.[30]

In the future, it will be steadily more difficult for nations to follow policies that attempt to exclude foreign influences or that assert total control by the government. That is not a prediction of the disappearance of all centralized power; it is only an assessment that, on balance, the pressures that new information technologies introduce work in that direction. The persistence, or even resurgence, of ethnic, religious, and national antagonisms that has been evident since 1989 in Eastern Europe and the republics of the former Soviet Union shows that, in any given case, many other social forces come into play, forces that can in fact prove to be more powerful than the liberalizing tendencies of information technology.

Global Financial Markets

A quite different example of the intensified international integration of national economies and societies, and one equally or more dependent on information technologies, is the emergence of global-scale financial markets. Their significance for national economic policy and for international economic relations, and their dependence on the capabilities of technology for their very existence, make them particularly useful for understanding just how, and how much, technological change serves to alter the economic elements of international affairs.

It takes only a brief glance at the parameters of the international financial marketplace to realize the extent to which financial markets have been transformed in the decade of the 1980s. Growth alone has been striking: between 1980 and 1986, trading in Eurobonds rose from $240 billion to

$3.5 trillion; Euroequity offerings of common and preferred stock went from $200 million in 1983 to $11.8 billion in 1986; by 1988, trading in foreign equities accounted for nearly 30 percent of total equity trading on the London Stock Exchange.[31] The U.S. Securities and Exchange Commission estimates that foreign investors carried out about $131 billion worth of transactions in U.S. stocks in the first half of 1986, more than three times the amount for all of 1982.[32]

Probably most telling is the comparison between foreign-currency trading and trade in goods and services. By 1986, trading in foreign currency was at a level of $330 billion per day, while trade in goods and services was under $6 billion per day—less than one-fiftieth as much.[33] A year later, the New York Bank Clearing House alone had a volume of trade in foreign currency in excess of $500 billion per day.[34]

Financial innovations are "running rampant," in the words of Joan Spero, executive vice president of the American Express Company. The capabilities of the new technologies allow an array of wholly new financial instruments. For example, currency and interest-rate swaps, which allow borrowers to trade differential access to currencies, amounted to $3 billion in 1982 but had reached $341 billion only four years later, in 1986.[35] Entirely new assets have been created out of assets previously untraded, such as mortgage packages. In 1987, those trades in mortgage packages equaled approximately one-half of the volume of new mortgages in the United States.[36]

Purchases of computer equipment by banks, security houses, and stock exchanges have also grown precipitously, with banks alone spending $30 billion per year by 1986— an investment increasing by 16 percent per year.[37] Computer capacity of the New York Stock Exchange was 100 million shares per day in 1978 and is expected to rise to 1 billion per day in the early 1990s.[38] Securities firms in the United States planned to spend $7.5 billion on computers in 1991, fully one-fifth of their total outlays that year.[39] Moreover, telecommunications have now made it feasible to trade instantaneously essentially anywhere in the world, so that a twenty-four-hour global securities market is now realizable, and probably irresistible.[40]

These numbers give some sense of the international dimensions of this explosion in the scale of activity, truly a matter for superlatives. Why did it come about, and what are its implications for governmental autonomy in economic affairs?

The single most important factor in making it feasible was the rapidly increasing capabilities of information technologies. At the most general level, the technologies made possible a continuing improvement in the ease and cost of transactions independent of distance, which made practicable large increases in the volume of transactions without reference to location. They also made it possible for traders to act on small and transient (on the order of seconds) differentials in exchange rates or prices between stock exchanges, contributing to both the velocity of money and the magnitude of capital flows. And traders could more easily develop innovative instruments and strategies to reduce the risks of volatility of prices and exchange rates. The volume and timeliness of information and the ability to carry out complex calculations quickly to determine risk and potential gain proved to be heady stimulants for innovation in financial instruments.

Furthermore, the ease of moving money and transactions meant that money managers could seek countries with favorable financial regulations, avoiding those with onerous rules.

Structural changes are accompanying this growth in markets, and changing technology is gradually erasing distinctions among banks, securities firms, and industrial lenders and stimulating the growth of international alliances, sometimes across traditional institutional lines (e.g., banks and security houses).[41] Large institutional investors are coming to dominate markets, further stimulating the development of innovative techniques to manage large pools of capital. Spero notes that British pension funds increased their foreign investment from $9.7 billion to $56.6 billion between 1980 and 1986; foreign investment by Japanese pension funds went from $400 million to $14.5 billion over the same period.[42]

This massive change in the scale of activity and in the structure of financial markets, has obvious implications for the autonomy of governments, and thus for international affairs, primarily by leading to constraints or limits on national freedom of action. The change forces each country's government, regulators, banks, brokerage houses, and industry to operate in a quite different framework and to accept the presence of nonnational actors with enhanced influence in the determination of policies and behavior. Economically powerful nations, such as the United States, may still have more influence on the international financial system than others, but even they can never return to the level of autonomy they believed (not always accurately) they had in the past.

But has the loss of national autonomy in economic and financial policy been as striking as might have been anticipated given the spectacular development of global financial markets? The sheer volume of financial flows, for example, could have been expected to have substantial implications. Spero argues that the flows "reduce the autonomy and effectiveness of national monetary policies" by making it "difficult for policymakers to define and interpret money and credit aggregates" and by exacerbating the volatility of exchange rates.[43] Others make similar arguments; to the nonspecialist, the effect of the volume of financial flows seems at first glance self-evident.[44]

Yet it is not self-evident. Economists note, for example, that the actual impact of massive currency trading on the world economy is far smaller than the volume would suggest. Over 85 percent of the trading is within financial organizations, representing short-term positions held for a matter of hours only and with little or no impact on trade in manufactures or other economic interests.[45] If governments attempt to tie their monetary policy to maintenance of a specific exchange rate in the market's transactions, then their policy will be hostage to the market. But if they make no such attempt—and the governments of all major economic powers characteristically do not—then their monetary policy is much less constrained by the scale and international nature of the market than might at first be expected. In fact, the global market can be an important aid *to* national economic-policy objectives. The expanded financial capabilities of that market, for example, were of material help to the United States in the 1980s, making it feasible to raise the funds to finance that nation's huge budget deficits.

Thus, the existence and size of the international market is relevant to the setting of national policies; but, as long as the market is running smoothly, it is not a materially larger factor in national policy-making than markets and international monetary transactions were in the past.

There is, of course, a catch: as long as the market is running smoothly. These markets depend fundamentally on the operation of large information systems, and this dependence (discussed in generic terms more fully later in this chapter) carries with it unavoidable risks of machine failure, of willful disruption, and of instabilities in internal system dynamics. Protective measures can be taken against external disruption or machine failure, though increased security necessarily means lower efficiency.

The problem of instabilities stemming from dynamics of the system is perhaps the most dangerous. Preprogrammed instructions for the computers at the heart of the system are necessary to cope with the high volume of transactions, the instant availability of information, and the opportunities for capitalizing on small differentials in rates. The possibility of sudden price swings encourages the use of preprogrammed instructions to enable the automatic sale or purchase of investments in case of a rapidly changing market. In the absence of adequate safeguards, these two factors, combined with the herd instinct that often characterizes security markets, create a system of high volatility. The market crises of October 1987 and October 1989 showed how tightly coupled and volatile the system has in fact become.[46]

Thus, the concern over the loss of national autonomy and authority and over the much-increased exposure to forces external to the nation comes primarily not from the regular operation of these massive new international markets, but rather from the vulnerability of the system to intentional or inadvertent failure or to the instability that grows out of the market's internal dynamics, either of which can have major economic consequences.

It might have been thought as well that the transformation of financial markets from a series of weakly interacting national markets to a globally integrated marketplace would mean that national differences in markets no longer exist or, if they do, are simply an anachronism that would not pose any dangers.

Any such conclusion would be, so far at least, a considerable exaggeration. National security markets, for example, appear to maintain internal dynamics of their own: trading rules and restrictions vary among markets, and local investment still dominates. A summary article in *The Economist* points out that the average correlation of stock prices among twenty-three of the largest markets was only 0.222 for the six years before the crash of October 1987.[47] Though it has risen about 50 percent since then, it is still far below what might be expected if there were, in effect, a single market. The global market, therefore, is less uniform than might be assumed; it is made up of nationally based elements that still preserve considerable variety and that operate with local rules and controls.

These differences in rules—for example, with respect to capital requirements for banks—create risks of failure of weak banks, which, in the tightly coupled system of international finance, could quickly propagate to bring the whole system down. Near-misses of that kind are not unknown.[48] Other existing incompatible regulations—disclosure, tax policies, and other forms of oversight—are also possible sources of instability in the system. Similar problems exist for brokers, leading the International Organization of Securities Commissions to consider new efforts to "harmonize the diverse rules that govern the international equities market" so as to avoid the danger of collapse of a brokerage house that could quickly send devastating shock waves to other markets.[49]

The international clearance and settlement system is a source of potential vulnerability as well. To quote the president of the Federal Reserve Bank of New York, E. Gerald Corrigan, "A major mechanical breakdown, liquidity problem or, even worse, default in one of these [clearing and settlement] systems has the potential to seriously and adversely affect all other direct and indirect participants in the system, even those that are far removed from the initial source of the problem."[50] These systems are overwhelmingly, and increasingly, technology-based, as are those for the trades themselves; that is the only way that operations of the required magnitude could be managed. Yet the meshing of the various national systems is exceedingly difficult, as they operate with different capabilities, settlement cycles (in 1991, the British still relied on the physical exchange of documents), operational standards, and financial strength.[51]

All these characteristics will be under increasing stress as the volume grows and the pressures of time become more severe.

Curiously, one of the major settlement systems—CHIPS (Clearing House Interbank Payments System)—was created by and is managed by private banks, rather than by a government. It is a measure of the degree to which financial markets, perhaps in the United States more than in other developed countries, are outside the direct management (as opposed to oversight) of government. This independence has not been a problem, so far at least, for governments determined to preserve control over their economies; all of the participants, public and private, have the same goal: smooth and reliable functioning of the market.

In sum, it appears that the more significant implication for international affairs of the creation of a global financial market lies in the danger of system vulnerability, rather than in the real but surprisingly limited loss of autonomy over domestic economic policies.

The dangers of volatility are well recognized by governments and by the banking community, though the development of appropriate counter-measures is time-consuming. The Standing Committee on Banking Regulations and Supervisory Practices (called the Cooke Committee, after its chairman) was formed in 1974 following the realization that traditional responses to recent bank failures were inadequate. The committee has been a forum for the development of agreed standards among the G-10 countries (the world's major financial powers—excluding, however, some of the rising banking centers, such as Singapore and Hong Kong), though its role in regime formation "is still in the early stage of development."[52] Its first major result was the Basle Concordat, which laid out a set of principles for handling banking crises; more recently, in 1987, it promulgated the Basle framework, which was to establish an agreed minimum level of capital for banks by 1992.[53]

Useful as these steps are, progress is slow and improvement partial. What may eventually be required is more than simply greater consultation and cooperation in the development of safeguards and compatible regulations. It has been suggested that much more ambitious international institutional innovation is needed. Spero, as a senior officer of a major credit institution, calls urgently for an international "lender of last resort" that would be committed to act in a crisis to protect the financial system and the economy from "unacceptable and unforeseen shocks"; W. Michael Blumenthal, former secretary of the U.S. Treasury and CEO of a large multinational company, asks, "For the joint management of our world financial market, some bold new thinking is especially needed. . . . Is a single world central bank as yet too visionary an idea?"[54]

Nations are surely far, as yet, from granting to an international institution the degree of autonomy that would be implied by the creation of a world central bank, though the magnitude of the global financial system and its vulnerabilities may have pushed the world considerably closer to that point. Even the European Community has found it difficult to move definitively toward a European central bank. Any such institutional innovation of a global scope, if it ever proved necessary or feasible, would certainly move the international political system a long way toward what would have to be seen as a fundamental structural change that would permanently alter the economic independence of nation-states.

So far, however, the emergence of global financial markets, made possible by advances in technology, can be said to have led to some quite new and more-extensive transnational and international economic relationships that have served to constrain the economic autonomy of governments—but not more than increased interdependence already has.

It is the inherent vulnerabilities of the massive, tightly coupled financial systems that carry the seeds of larger change in international economic structure and relationships. And those vulnerabilities are closely related to the characteristics of the technologies that made the systems possible in the first place. As long as the systems run reasonably smoothly, major structural change will come slowly and with difficulty; the political and economic barriers to major change discourage anticipatory action that would be both expensive and not easily demonstrated to be necessary. If adequate anticipatory changes are not made, it is eventual catastrophic breakdown that will bring about more structural change. The extent of the breakdown would then determine just how far restructuring would be necessary or feasible and thus what level of cooperation and constraints on national autonomy would result.

Multinational Corporations

One of the more evident, and contentious, results of the integration of economies has been the growth of the size and economic power of multinational industrial corporations. Their growth is not simply a result of technological change; but such change helped to create the economic incentives for activities on an international level. Lower production costs and increased capacity, resulting from advances in transportation and information technologies, stimulated the dispersion of production on a global scale, the entry of new countries into the international marketplace, and the development of a world market for goods and services. Increased economies of scale and higher r/d costs that mandated larger production runs contributed to the growth of firms large enough to take advantage of international production and markets. And the diffusion of technological competence worldwide made it important to be in a position to tap technological developments wherever they occurred.

A debate has long raged over the political and economic significance of these multinationals, with no clear resolution. Some observers, such as Richard Barnet and Ronald Müller, believe they are the most important, and dangerous, institutional development of this century: "The global corporation is the most powerful human organization yet devised for colonizing the future. . . . it has become an institution of unique power."[55] They argue that multinationals are amassing political power and establishing political legitimacy as they gather more public decisions into private hands in a situation in which "government as a practical matter can no longer control them."[56] Others see them as important new elements on the international political scene, able to exert economic and even political power, but with complex and very much more mixed effects on home or host nations.[57]

Either way, the growth of multinationals represents an example of how technological change has contributed to the evolution of an important element in international affairs, in effect making possible and even stimulating the creation of new sources of power that necessarily affect relations among nations. If the more draconian view is correct, these corporations could in time bring about a fundamental reordering of the structure of the international system, overwhelming the power of the nation-state. There is little that has happened since Barnet and Muller made their arguments, in 1974, to support that thesis. In fact, to some extent the debate has turned, as multinational corporations have in some circumstances become key elements in *national* strategies to maintain international competitiveness.[58]

A more restrained view of the impact (at least to date) of multinational corporations is that their pervasive international activities, and their relative ability (whether exercised or not) to evade specific national policies or to act independently of the policies of their home government, simply constitute additional constraints, among many such, on the freedom of action of national governments to control the formulation and implementation of national policy.

Whatever view is taken of the current debate, the relevant question for our inquiry is whether the continuing scientific and technological advances that originally stimulated the growth of multinationals will in the future contribute to movement in one direction or the other—whether these advances will necessarily continue to increase the incentives for international industrial operations and structure, thus contributing to the growing scale and, implicitly, power of the corporations. Is technological change neutral in this respect, or is there a natural bias in the characteristics of technological change that, on balance, is more likely to strengthen economic power based in institutions outside governments, in effect adding to the constraints on the economic autonomy and authority of governments?

A complete analysis of that question would require more attention to that one issue than is appropriate. It will be sufficient for our purposes to examine one crucial factor in the development of multinationals: the flexibility offered by technology that makes dispersion of production economically feasible and attractive. That is a sufficiently important factor to serve as a representative of the many others that influence the development of multinationals.

The profitable dispersal of industrial production on an international scale was made possible by dramatic advances in transportation and information technologies that allowed rapid and reliable movement of goods and communication at low cost. By taking advantage of differential cost factors and regulations, of economies of scale, and of decreasing tariff barriers, corporations could maximize the efficiency of production. Production could now be located wherever it was most economically advantageous, independently of the market to be served or even of where finished goods were to be assembled. Though dispersed production is a phenomenon only of recent decades, by 1983 it had reached a level at which "more than half of U.S. sales of certain products in textiles and electronics are assembled abroad. . . . The imports of products assembled abroad have reached a level of almost a sixth of total U.S. imports of manufactures and about a quarter of imports of manufactures from developing countries."[59] In 1986, sales of high-technology products by the *foreign* affiliates of U.S. companies reached a level *twice* as large as U.S. high-technology exports.[60] At the end of the 1980s, multinational corporations employed some 45 million people worldwide and accounted for over 90 percent of all U.S. foreign trade.[61]

Though many nontechnical factors were relevant, it was technological change that made global production feasible and economically attractive. Thus, it could be said that the outcomes of the scientific and technological enterprises have tended on balance to support the ability of multinationals to disperse production on a global scale and in the process have implicitly contributed to their expansion and decreased the ability of individual governments to influence or control their activities.

But will that always be true? Another characteristic outcome of science and technology is to provide more options and the attendant flexibility for the achievement of corporate objectives. The pressures and incentives in the development of technology that have resulted in the dramatic improvements in transportation and communications technologies already realized will continue to lead to change that will again alter the incentives and the economics of dispersed production.[62] In fact, Yves Doz and others conclude: the application of new technology has so lowered manufacturing costs that other nontechnical factors have become more important in multinationals' decisions on siting or expansion; some new technologies may actually reverse economies of scale or allow greater flexibility of production, reducing the need for large plants; "just-in-time" manufacturing concepts (in which suppliers are tightly coordinated with production needs to avoid inventory stocks) work best with co-location of assembly and suppliers; and the labor-cost advantages of low-wage locations become less significant as production technology becomes automated and more capital-intensive.[63]

Thus, technological developments, which contributed to the incentives for globalization of production, now contribute to the moderation or even reversal of those incentives. The result is more diversity in industrial patterns and more diversity in the ways governments can influence the economic choices that firms make. In effect, rather than leading to developments that point largely in one direction (expanded globalization of firms and increasing impotence of governments), technological outcomes are sufficiently neutral as to make nontechnical factors much more important in the actual economic outcomes.[64]

Harvey Brooks and Bruce Guile, in a National Academy of Engineering report, sum it up well:

> The trend during the last 30 years has been toward global homogenization of markets and transnational integration of production. Yet there are signs of the emergence of countervailing pressures resulting from technological, managerial, and political developments that appear to be giving a competitive advantage to more localized production and distribution. . . . Because modern technologies are both flexible and diverse, other factors may be more important than technology as a determinant of organizational structure.[65]

Doz makes two other observations about limitations on globalization, both supporting the argument that nontechnical factors, including political factors, increase in importance as technology introduces more flexibility. One is that government policies need not be impotent, as they are often said to be; they may, in fact, dominate the factors that determine company policies in this global-production dimension. He cites in particular the rise of protectionist tendencies, which can greatly alter the strategic planning, operation, and profitability of multinationals.[66]

Doz's other observation is that as companies' international operations and production become increasingly characterized by diversity, the managerial requirements become extremely demanding; there are likely to be few companies with the capacity to meet those conditions fully. Large multinational companies may as a result become less successful concentrations of power and thus less, not more, formidable competitors of governments.

Of course, government policy with respect to multinational corporations will not necessarily be effective in maximizing national economic objectives. The contentious political debate over the dangers of foreign direct investment in the United States, for example, is often overshadowed by protectionist policies that impose large, "deadweight" costs that outweigh whatever costs the multinationals themselves might levy on the economy.[67]

Thus, although technological change has been a crucial enabling factor in the rise of multinationals, there is much reason to question whether continuing technological change will necessarily encourage their further expansion—expansion that would enhance their economic power vis-à-vis governments and their role in the evolution of international affairs. The increasing flexibility that technology provides makes factors other than technology much more crucial in that equation; in effect, technology in this case becomes more neutral in its significance as the characteristics of technological outcomes continue to change. In fact, in a world increasingly concerned with economic competition, by far the more important issue is the role of technology, and of multinational corporations, in the competitive capacity of nations—a subject to which we will now turn.

Economic Growth, Trade, and Competition

Science and technology have been important elements in economic growth, and hence in international power and influence, since at least the early 1800s. Simon Kuznets said it simply: "The epochal innovation that distinguishes the modern economic epoch is the extended application of science to problems of economic production. . . . certainly since the second half of the nineteenth century, the major source of economic growth in the developed countries has been science-based technology."[68] In chapter 2, we traced the evolution of the intensifying web of relationships between the scientific and technological enterprises on the one hand and industry and national governments on the other, and the growing importance of their interaction to both. Many economists and industrialists could be enlisted to make the point, but it hardly seems necessary; the significance of science and technology for the economic growth of the last two centuries is not, in broad terms, an issue.[69]

Curiously, just *how* significant they are *is* an issue. Ever since Robert Solow's pioneering 1957 paper that found the increase in output for the American nonfarm economy during 1909–49 to be dominantly a result of technological change, economists and others have been attempting to develop better estimates of the economic return on r/d investment by firms or government.[70] The results for private-sector investment are suggestive of high rates of return: Edwin Mansfield, summarizing a number of studies by economists, reported rates of return to r/d in manufacturing industries to be about 30 percent. The median "social rate of return" (the return to the society as a whole rather than to the firm making the investment) in Mansfield's own studies of specific innovations varied from 56 percent to 90 percent, in each case at least double the private rate of return to the innovator.[71]

These are substantial estimates, considered by the authors to be conservative but nonetheless treated cautiously both by them and by those attempting to use the results for allocations of funds for r/d. There are several reasons for the caution: the complexity of the relationships, which makes the problem of ascribing clear causal patterns or of identifying truly independent variables particularly difficult; the uncertainty in the measurement of both inputs and outputs; the problem of accounting for qualitative as well as quantitative advance; the difficulty of separating the effects of technological innovation and change from those of r/d; and the fact that much of the r/d expenditures of some countries is devoted to military and space objectives whose contribution to the economy is not easily assessed.

Notwithstanding these problems of measurement, there is no doubt of the general and widespread acceptance of the key role of science and technology in economic growth today.

A nation's economic strength is not simply a matter of economic growth, however, nor is its capability in science and technology the only determinant of growth. Many other variables intervene and are important in their own right, including, *inter alia*, the management of the transition from scientific and technological strength to commercial success in the marketplace. But the growing proportion of economic activity that involves technology-based goods and services, the crucial dependence of economic performance on technological contributions to productivity, the expanded proportion of high-technology trade in the national economy, and the international economic competition among states increasingly competent in technology now make the scientific and technological capabilities of a nation one of the fundamental elements in its economic position. They are a necessary, if not a sufficient, condition for economic strength.

There has thus been an important evolution in one of the primary variables in international affairs, as scientific and technological capability has become a major determinant of the relative economic status of nations. The significance of that variable is not tied to any specific technology or technological change, but is a product of the entire scientific and technological enterprise, of how that enterprise is coupled to the rest of the economic and political structure of a country and, in particular, of how effectively an economy translates technological advance into improvements in national productivity. Technological strength has become an

essentially new, or at least newly important, aspect of geopolitical assessments, complementing or replacing more-traditional measures of economic power, such as resource endowments, basic industrial capacity, or
population size.[72]

The recognition of this key economic role of science and technology has
become a standard in the rhetoric of political and industrial leaders; a
quotation from the French statesman Raymond Barre, at a conference of
leading figures in industry and politics, is representative of many: "New
growth would be quite different from the past. It would be led by scientific and technological innovation, meaning that economic capacity among
nations would in future be won by those with most knowledge. A nation's
most jealously guarded accounts would be its education skills and its investment in science and research."[73]

The new significance of technological capability to a nation's economic
strength has had important repercussions along all the axes of international affairs: among market-based industrialized economies, between
market economies and centrally planned economies, and between developed and developing countries. We will explore the role of technological
capability in all three settings, considering first industrial market economies and then command economies in the context of growth, trade, and
competition. The role of technological capability in the relations between
industrial and developing nations, which raises issues of a somewhat different kind, will be considered separately in the context of North-South
transfer of technology and dependency.

Industrial Market Economies

A clear indication of the role now played by technological capability is
the emergence of high-technology competition as a major factor in economic relations among industrialized market economies. Its rise in significance paralleled the decline of postwar American dominance of international economic affairs as the industrial economies of Europe and Japan
recovered from the devastation of the war, helped along by substantial
support from the United States. American technological leadership, translated into successful innovation in the marketplace, became a major
asset in the U.S. economic arsenal as advantages in other areas began to
fade. But the value of that asset has been called into question as other
countries, particularly Japan, challenge America in capacity for innovation and prove adept in the competition for markets, higher productivity,
and the acquisition of advanced technologies. Though the U. S. remains
the overall leader in technological competence in the early 1990s, there
are serious concerns about its long-run competitiveness in the innovations
and productivity improvements that should flow from that leadership.[74]

The competition grows steadily more demanding for the United States; Japan poses strident challenges and Europe looms as a potentially more powerful competitor after its reduction of internal barriers planned for the end of 1992. The very determinants of success in technological competitiveness may be changing, as rapid, incremental product improvement, rather than technological leaps forward, may be the more important defining characteristic of a nation's competitive position.[75] At the same time, existing arenas of competition expand and new ones emerge: supercomputers, computer memory chips (DRAMs), materials technologies, high-definition television (HDTV), integrated services digital network (ISDN), biotechnology, and possibly military and commercial aircraft, among others; most of these represent large economic stakes for the countries involved. It is not too strong to say that the outcomes of these contests will determine the economic status of nations in the international system for many years in the future.

HIGH-TECHNOLOGY TRADE

Trade in goods and services is the stage on which international economic competition primarily takes place. The rules of the game are controversial and in flux, but any trading relationship implies a degree of dependence of one country on another. In a schematic description of a liberal trading system, mutual dependency of trading partners is inherent in the system, with each country exploiting its comparative advantage to export what it can produce at lower cost and importing products and materials that it cannot produce as inexpensively as others. When the system is working properly, all nations will benefit from what is a positive-sum game.[76]

Perhaps the most significant effect of technological change on the concepts that underlie a liberal trading system is the alteration in the content of comparative advantage. In traditional economic theory, the sources of comparative advantage, which determines the flow of trade when the trading system is open and unfettered, were the relative endowments of a nation in the factors of production: natural resources, agricultural land, labor, and capital. Now, comparative advantage, or what Michael Porter calls competitive advantage, lies in a broader set of characteristics that prominently include an economy's capacity for technological innovation to improve productivity. As a result, comparative (or competitive) advantage can be "created"; that is, it flows from what an economy can produce through its human skills, its organization, and the competence and productivity of its scientific and technological base. It stems, in other words, from national policy and corporate decisions, rather than from natural endowment.[77] Though the malleability of comparative advantage is not a wholly new development—a nation's ability to improve its human and

capital resources was always a relevant factor—it is an appreciable alteration of one of the standard geopolitical measures of a nation's international
status and position.

One measure of technological competitiveness is performance in trade
in high-technology products, which constitute a growing share of the
trade in manufactured goods: from 1970 to 1986, the proportion of technology-intensive goods grew from 16 percent to 22 percent of world manufactured exports, reaching 37 percent of the United States' manufactured
exports and 33 percent of Japan's in 1986.[78] This expansion of high-technology trade should be seen in the context of greatly increased penetra-
tion of foreign trade as a whole in the U.S. economy, by far the largest in
the world. In that nation, which among major Western countries had been
the least involved in international trade, exports and imports together
increased from about 10 percent of the economy in 1950 to about 25 percent in 1987.[79]

But at the same time, the U.S. position in world high-technology trade
fell materially from its postwar high. Of the total world trade in technology-intensive products, the U.S. share fell from 27.5 percent to 20.6
percent in the two decades between 1965 and 1987. During the same
period, the Japanese share grew almost threefold, from 7.2 percent to
18.9 percent; in 1989, Japan exported more than five times as much as
it imported in high-technology products.[80] Moreover, the United States'
large positive balance in high-technology trade, which reached as much
as $27.4 billion in 1981, turned negative in 1986 by $2.6 billion, recover-
ing modestly in 1987 and 1988. Its deficit in high-technology trade with
Japan was over $22.3 billion in 1988, though it has been able to main-
tain a positive balance with the European Community.[81] Since 1988, a
combination of productivity improvements and a weakening dollar have
improved the picture for the United States balance with Japan, though
the long-term outlook remains uncertain at best.[82] It has been tempting
for many to see in these statistics permanent decline for the United
States; but, as Joseph Nye points out, these reflect to a considerable
extent what he calls the "vanishing World War II effect," which put the
United States in an unusually favorable position for a number of years
after 1945 because of the war's devastation of its competitors.[83] Now
that the aberrations of the war have worked their way through the system, he argues that a more "normal" distribution of strengths is in evidence, with the United States still the overall leader and capable of
remaining so if it recognizes adequately the changing nature of the challenges it faces.

It is evident that, whether or not a "normal" distribution has been
reached, the competitive position of industrialized countries in high technology, as measured by trade performance, has been greatly modified

since the 1960s. Japan has become a major exporter of high-technology products, challenging the U.S. technological leadership successfully in many significant commercial areas and reaping substantial economic gains. Europe has yet to alter its position appreciably or to show what its capabilities may be, particularly in the post-1992 market.

This new competitive situation has led to considerable soul-searching as nations have attempted to find the appropriate policies to improve their technological performance. The United States has been particularly occupied with the question, at least at the level of debate, with a mixture of chagrin, anger, and some determination as the country sees what it considers its traditional technological birthright being seriously eroded by Japan. European nations have also been singly and collectively attempting to mobilize more effectively in anticipation of the 1992 moves toward increased economic integration.[84]

A major difficulty is that, though the importance of improving competitive technological position may be clear, how to do it—especially in those economies apparently losing their competitive edge—is not. Proposed measures are controversial because they involve political and economic interests of a nation that go far beyond technological matters alone, and far beyond international affairs alone. The debate in the United States has tended to focus on whether the government should intervene in the economy by selecting commercially important technological fields for special support. The issue is usually seen in the context of controversies over "industrial policy," with strong views expressed that reflect deeply held ideological, economic, or political convictions.[85]

Some support Michael Porter's general position that government should not intervene directly in industry, but should help in creating the essential home environment—including education, research, infrastructure, and capital availability—that will spawn creative, innovative companies.[86] In his view, companies are basically national in orientation, whatever the level of their international activities and ties, and are crucially dependent on their home environment and particularly on its internal vigor and competitiveness. Others call for an industrial policy that would involve varying degrees of intervention in the economy, with special measures to help improve the technological performance of economically significant high-technology industries.[87]

Advocates of greater government intervention have received support from emerging, and controversial, ideas in economic theory. Proponents of what has come to be called strategic-trade theory argue that there are "strategic" sectors in an economy—high technology being a prominent example—that receive a higher return to investment or generate special benefits for a society that are not reflected in the prices paid to producers. Thus, contrary to the traditional assumption that market prices are the

sole appropriate guide to the allocation of resources, the positive externalities of those sectors for society as a whole would justify biases in government policy in their favor. Paul Krugman sums up the argument with
regard to high-technology industry:

> Because of the important roles now being given to economies of scale, advan-
> tages of experience, and innovation as explanations of trading patterns, it seems
> more likely that rent will not be fully competed away—that is, that labor or
> capital will sometimes earn significantly higher returns in some industries
> than in others. Because of the increased role of technological competition, it
> has become more plausible to argue that certain sectors yield important exter-
> nal economies, so producers are not in fact paid the full social value of their
> production.
>
> What all this means is that the extreme pro-free-trade position—that markets
> work so well that they cannot be improved on—has become untenable. In this
> sense the new approaches to international trade provide a potential rationale for
> a turn by the United States toward a more activist trade policy.[88]

It may be reasonable to accept the idea of such strategic sectors; but, as
Krugman notes, that does not mean it is obvious how those sectors can be
identified or what policies ought then to be followed. In fact, however, all
national governments have long been providing explicit or implicit support of one kind or other for their high-technology commercial industries.
Whether it be direct government intervention of varying kinds (including
protection) in Japan or indirect support through defense r/d in the United
States, all governments have had de facto industrial policies.[89]

Those policies, however, have been considerably more comprehensive
in most other industrial countries than in the United States. Notwithstanding the extensive support in that country for basic science and for
large-scale r/d to serve the public objectives of defense, space, agriculture, and public health, the U.S. government has been unwilling to make
a substantial commitment to use public funds to advance commercial technologies. Other advanced industrial nations that are the chief competitors
of the United States, notably Japan and Germany, have been quite willing
to do so, providing support for commercially oriented r/d, for the diffusion
of technology, and for strengthening industrial capacity for the effective
absorption and development of commercial technology.[90]

The United States has finally begun to relax its strictures against greater
involvement in industrial or technology policy. Enabling legislation was
passed in the 1980s to permit creation of research consortiums of private
companies without violation of antitrust laws, and to establish an Advanced Technology Program in the National Institute of Science and
Technology (NIST), formerly the National Bureau of Standards, in the
Department of Commerce. In 1990, what may prove to be a breakthrough

White House statement on technology policy was released by the Bush administration.[91] The statement affirms the government's willingness to cooperate with industry in the development of generic, enabling technologies. The earlier legislation has led to a small but innovative program in NIST and to the formation of many research consortiums; in the case of Sematech, a consortium to develop semiconductor memory chips, government funds have been provided for one-half the budget.[92] Other proposals, however, such as a consortium for the development of high-definition television and another for computer memory chips (U.S. Memories), have received only limited U.S. government support, and the latter failed even to come into existence.[93]

But allowing and supporting generic, precompetitive research is only the beginning of the task for the United States. The capability for competitive technological innovation also requires attention to many domestic issues, such as human-resource development, the strength of educational institutions, the quality of the decisions for support of r/d, the effectiveness of tax incentives, the soundness of the capability for monitoring and using technological knowledge from abroad (a particular problem for American engineers, who until relatively recently had little reason to follow developments in other countries), and many others.

It should be noted that the relevance of these policy areas—areas commonly thought of as domestic issues—to a nation's technological competitiveness has served to greatly expand the substance of trade negotiations, as the issues relevant to technological trade can no longer be confined to matters such as tariffs and direct export subsidies. National policies and practices in areas that previously received little attention in trade negotiations are now directly relevant. In effect, the structural differences among nations become, in Laura Tyson's terms, a major element of trade conflict in technology-intensive industries.[94] As David Mowery and Nathan Rosenberg observe, "Because of their effects on trade flows, domestic subsidies for research or investment, government procurement, intellectual property regimes, patent policies, regional development policies, and other policies that historically have received little scrutiny from trade policymakers are now central to trade negotiations."[95] Other subjects become relevant as well, even the informal relationships of the business, government, and financial communities.[96]

Thus, the role of technology in international trade and competitiveness has meant that a set of subtle and particularly difficult issues has become a necessary part of international trade negotiations. New rules, new agreements, and perhaps new kinds of relationships that reach deeply into national styles and cultures will be necessary to maintain a "level playing field."[97] Substantial agreements to establish workable trading rules may be unreachable in that situation; or, to put it another way, if agreements are to be reached, they will require substantial compromises on sensitive domestic matters, and correspondingly will lead to significant change within nations.

It is also relevant to observe that in this context, basic science has become a more immediate element in a nation's international position. The greater scientific content of some technologies means that basic research that might be related to commercially strategic fields comes closer than ever before to being an instrument of international competition. That could be an advantage for the United States, still the strongest in basic research. But the growing economic relevance of research also increases pressures to limit the free dissemination of research results and to constrain the traditional openness of American university laboratories, where most basic research is performed.[98] Suddenly, principles of open communication, deemed by scientists to be so fundamental to the progress and health of science, are jeopardized by the close relationship of science to the economic competitiveness of nations.

And the more immediate economic relevance of basic research puts into bolder relief the consequences of possible national changes in attitudes and policies toward science. It appears that the United States in particular is facing new political challenges to the science policies that have prevailed since the end of World War II. The combination of a divisive new populism that questions allocating funds on the basis of excellence; the end of the cold war, which removes the security justification for large funding and for minimizing politics in allocation decisions; a continuing fragmentation of national politics, which diffuses central authority; and an apparent growing public wariness of the fruits of science and technology could well result in a more diffuse, smaller, and less productive policy toward science and technology in the United States.[99] The implications of such a situation, if it all comes to pass, for the international trading system and particularly for the place of the United States in it, are not hard to imagine.

CRITICAL TECHNOLOGICAL DEPENDENCIES

The competition in high-technology products and processes that will be prime determinants of the future economic status of nations might have another feature as well: it could lead to virulent technological dependencies between industrial countries that would have serious economic and political consequences. Such a dependency would result if a vital or "critical" technology—one essential for the design and production of other high-value technological products—was of such a nature that competence in it, once lost, could not be regained at an acceptable level of cost and quality relative to the competitor who set the pace in the technology. A technology of that kind would clearly fall into the category of strategic technologies discussed above, but its unusual importance to an economy justifies the more compelling idea of criticality.

Such dependencies are not uncommon for small or medium-size nations that necessarily must rely on others for many essential products. A major economic power, however, would be less than comfortable with the political and economic costs of the vulnerability that could exist if the control of a technology were concentrated in another country able and willing to use the leverage it provided. It would be a classic monopoly situation—never a comfortable one for a major economic power that doesn't hold the monopoly, but made more serious because it involves a vital commodity with potential effects reaching far beyond the technology itself.[100]

Semiconductor chips—the small silicon wafers on which complete electronic circuits have been constructed—have become the crucial component of computers and of an increasing array of information and consumer technologies, and may be just such a critical technology. They are the ubiquitous and indispensible components of a growing proportion of the products of modern economies; to produce them at levels of competitive performance, quality, and cost, while staying abreast of the incredible pace of their development, requires many capabilities: high-quality research at the frontier of many fields of science and technology; superior competence in an extremely exacting and constantly evolving manufacturing process (involving working with circuit elements at submicron dimensions—less than a millionth of a meter); technical information that is derived from intimate contact with user as well as designer communities and with related scientific and technological research; and familiarity with know-how, or embedded knowledge, in manufacturing that can only come from experience at the manufacturing and research frontier. Nor can the knowledge required to manufacture a chip with acceptable quality and cost be reverse engineered from the chip itself.[101]

As Michael Borrus, Laura Tyson, and John Zysman, in their study of international trade in the semiconductor industry, point out: "it is often the case that potential users of a new technology require knowledge of products in development months and even years before such products are available on the market if their own research and innovation activity is to be successful. Often the only way to acquire such information is to be involved actively in related research areas and to participate in the related scientific communities."[102] Thus, the user communities must be close to, and have frequent interaction with, those designing and fabricating the chips. A senior Japanese engineer in the Sony Corporation in charge of the miniaturization of video cameras put it succinctly: "You cannot order the parts and hope to put something together. If you don't know how to make the parts, you don't know how to make the camera."[103] And the president of Toshiba observes that "only if you are good and fruitful in a previous generation can you afford to go ahead in the next stage."[104]

In sum, the process of producing chips of competitive quality and performance is unusually demanding and is evolving rapidly; it would be extremely difficult, or unrealistic, to attempt to recreate the ability to produce them, once lost. In turn, the ability to develop and design competitive, downstream products that employ new generations of chips requires interaction between the downstream designers and the development process for the chips themselves. As a result, the inability to retain a viable semiconductor industry by a nation endeavoring to remain competitive in high-technology products—an inability that might come about from poor performance caused by any of a number of possible factors, including aggressive and unfair competition—could have repercussions far beyond the economic loss of that industry alone. To quote Borrus, Tyson, and Zysman once again:

> Under these circumstances, if the United States loses a substantial portion of its semiconductor industry to foreign competitors, it may lose the domestic scientific community on which the ability to innovate in semiconductors and related electronics industries depends. . . . In critical sectors, like semiconductors, on which the competitive positions of numerous other activities depend, a country's gain or loss in competitive position can result in a cumulative gain or loss across a whole spectrum of connected industries.[105]

Thus, it is not simply that a transnational dependency on semiconductor chips might be exploited by denying their sale at crucial times or by sharply raising prices, which would be the normal concerns in traditional monopoly dependencies.[106] Rather, the concern is about more-fundamental consequences that would be of substantial scale and persistence for many other industries important to an economy. And there could be repercussions as well in the security sector, where there is increasing dependence on commercial information technologies.[107]

This is not seen as an abstract argument in the United States. The U.S.-based semiconductor industry has encountered problems in producing DRAMs (dynamic random-access memory chips, key chips used in all computers) in competition with Japanese producers; those difficulties have led to a substantial loss of American sales and production potential in that industry, raising exactly the concerns outlined.[108] In the early 1980s, the United States dominated the industry, with more than 60 percent of the memory-chip market; for the early 1990s, the U.S. share is forecast to fall below 30 percent, with the Japanese capturing more than half the market and seven of the top market positions.[109] In 1988, for example, Japanese firms accounted for 95 percent of the supply of one-megabit memory chips (with one million elements on a single silicon wafer).[110] U.S. producers gained market share in 1990 for the first time in a decade, but the continued larger investment by Japanese firms is expected to result in a return to a growing Japanese share of the market.[111] The weakening of the Japanese, American, and European economies in the early 1990s, and the increased investment required to produce new generations of chips, has, however, reduced sharply the profits the Japanese can expect from their investment in chip-production technology.[112]

The financial stake is very large: Texas Instruments estimates sales of sixty-four-megabit chips, likely to enter the market in the mid-1990s or earlier, will be some $50 billion by the late 1990s, with the sales of products containing the chips possibly reaching $2 trillion.[113]

Some of the disquiet is focused not on the chips themselves, but on the equipment necessary for their manufacture, a technology also evolving rapidly. Easily available local knowledge and experience of the state of the art in manufacturing technology is itself of critical importance to the ability to stay abreast in the design and production of the chips. There need not be overt denial of information or of equipment by manufacturers in other countries; the lack of domestic competence at the frontier of the technology can be enough to lead to gradual loss of position.

Unease in the United States about the dangers of growing dependence on Japan for memory chips led to several policy moves. One was to negotiate an agreement with the Japanese that would tie the price of Japanese chips sold in the United States to the cost of their production and that would earmark a share of the Japanese market for U.S. producers.[114] This was intended as an antidumping measure, motivated by the belief that Japanese companies had artificially reduced their prices in the United States as a way of driving out American competitors and capturing market share. The agreement was intended to raise the price of chips, with the hope that that would encourage new investment in the United States and thus bring American manufacturers back into chip production. The immediate effects were to create a shortage of chips, drive prices up, and deliver huge profits for Japanese chip makers.[115]

The dismay over the extent of the possible dependency also led to the formation in 1987 of Sematech, an association of American firms whose purpose was to conduct the semiconductor r/d required to remain competitive with the Japanese or any other producer.[116] The security dimension of the issue provided a rationale for Department of Defense assistance of $100 million annually, intended to be one-half the Sematech budget. The assistance was essential because industry was unwilling to foot the entire bill, while, in the budgetary and ideological climate of the time, no civilian agency of government had the necessary authority or resources. Nor was there acceptance in the Reagan administration of the appropriateness of using public funds even for precompetitive development of a commercial technology; a security rationale was required. An attempt was also made to create a cooperative organization to develop technology for production of more-advanced chips—an even more unusual departure under American antitrust law. This was to be a joint venture, called U.S. Memories, among American companies. It was abandoned when the price of chips declined and not enough companies were willing to invest the necessary capital.[117]

The Europeans have entered the competition with their own, larger research consortium (with an expected $3–4 billion budget in the first half of the 1990s), JESSI (Joint European Submicron Silicon Initiative); five European governments and three large electronics firms will concentrate on advanced memory chips with circuit elements of half a micron or less.[118] The Japanese are mounting not one or a few, but perhaps a dozen programs to develop future chip-fabrication technologies, while the top five Japanese semiconductor firms are estimated to spend twice as much on r/d as their U.S. counterparts.[119] The Japanese also have their own consortium, Sortec—made up of thirteen large electronics firms—intended to contribute to the development of X-ray lithography.[120] That is the likely technology for etching chips with dimensions of one-quarter micron or less, a technology that will be needed for future generations of chips. Cooperation among companies is valuable because of the very high costs of r/d as the technology moves to higher-capacity chips and thus finer line sizes. It is noteworthy that the majority of funding for Sortec comes from Japanese industry, with the government contributing only $100 million over ten years. American industry has been willing to put up only one-half of the costs of Sematech, which therefore requires government funding of $100 million each year.[121]

Concern about the implications for a nation's economy of dependence on foreign sources for semiconductors is widespread; the arguments that lead to the concern are persuasive, persuasive enough to have caused departures from previous practice by the U.S. government (though rather limited in scale) and by American firms. There is no doubt that permanent loss of the capability to produce chips at acceptable quality and cost would have important economic, and perhaps political, repercussions that would be much broader in impact than the size of the industry alone would indicate.

Is there, in fact, an inherent point of no return, beyond which recovery from the loss of the technological capability is not possible in any reasonable time frame? Does the technology raise genuinely new or different issues than those from historical examples of international dependency?

The answers to those questions are not as certain as might be supposed. As a minimum, it is evident that as technology grows in importance in national economic performance, any unequal or uneven technological capacities among nations will be translated into dependencies of one economy on another. These are likely to be shifting, constantly changing relationships as nations struggle to improve their relative technological positions over time. Inadequate policies, whether on the part of government or industry, may doom countries to gradually worsening comparative performance and thus to increasing dependency. Such a dependency of itself would not be a novel situation, except that the focus would be technology—a newly important factor, but not conceptually different from other factors in the past.

Is there a point of no return for the ability to manufacture semiconductor chips that would mark a much more significant evolution in the meaning of dependency? The evidence is ambiguous. IBM and AT&T, which make their own chips in the United States and do not market them, are roughly equal to Japanese producers in technological capability. [122] In addition, the United States has been able, at least so far, to maintain its lead in microprocessors (chips designed for operational functions that are the basis of the operating systems in computers and computer applications), a lead that will serve to retain at least some of the competence that would otherwise be lost in a demise of U.S. DRAM production. [123] In fact, new developments in memory-chip design that customizes them for particular applications will make the market for such chips look more like that for microprocessors, playing to the United States' relative strength. [124] Furthermore, the increasing presence of Japanese firms in the American market, and cross-national research and production arrangements (e.g., IBM/Siemens or Motorola/Toshiba), will change the meaning of "domestic production." [125] Multinational firms may still have recognizable national identities, even if their operations are heavily international; but their activities as manufacturers of a critical technology cannot avoid making that technology available in at least a limited way to the surrounding economy, wherever the manufacturing takes place. Moreover, if those firms carry out r/d in the major markets in which they participate, there will be at least some local spin-off. [126] Whether these factors will prevent the feared loss of the technological capability may be problematic, but it is hard to be certain that they would not.

Many of the critics of U.S. policy, those who are raising the strongest cries of alarm, also argue that there is simply nothing resembling free trade in the international competition in semiconductors. Subsidies, protection, and predatory practices are seen to be rife; the traditional correctives of a liberal trading system are not applicable. Michael Borrus argues that market competition in semiconductor chips has turned into a "strategic game," in which the American industry—fragmented, ill-financed, and neglected by its government—has little chance against the vertically organized, well-financed, government-favored Japanese giants arrayed against it. [127]

The competitive situation in the semiconductor industry is complex, controversial, and certainly not yet proven to be critically different from more-traditional forms of dependency; but the potential significance of this new form of dependence for the economic status of nations is disturbing. The cries of alarm may be, and probably are, exaggerated; but whether they are or not, the challenge posed for any nation determined to compete in these technologies requires the willingness and the ability to confront central and sensitive policy issues in the attempt.

INTELLECTUAL-PROPERTY RIGHTS

Lastly, the matter of intellectual-property rights—a trade issue that has become prominent as a result of the formidable presence of technology-intensive products in international trade—deserves special note. Though the issue does not have such broad significance for a country's economic position as technological dependencies, it has become an important question on the trade agenda, and it neatly demonstrates how technological change can rapidly transform old issues into complex new subjects with ramifications across several major international concerns. In a nutshell, the problem is how to derive economic value from the promotion of innovation if property rights are not adequately protected when technology is diffused. The issue may be able to be resolved, with difficulty, within a nation; but there are widely differing laws and attitudes toward such rights among the nations of the world.

The issues of particular concern arise with respect to high-technology products such as computer software and chip designs, videotapes, genetically engineered products, and pharmaceuticals in general, as well as more-traditional items such as books and movies. The significant changes from the past lie in the economic importance of the property, the ease of copying and alteration, and the differing rules and practices of intellectual-property protections (patents, copyrights, and trade secrets) among countries. These products are now the basis of whole industries and can be major factors in determining a country's international economic status.

The nations with the largest stake in knowledge-based products are naturally anxious to be sure their products are protected, and not used or copied without recompense. Investments in developing and bringing to market new drugs, computer programs, or innovative chip designs require the sale of the product to repay the investment. But knowledge-based products are not like other technologies or, for that matter, other resources. For many products, copying presents no difficulties whatever;

sometimes, as for computer software, the capability is inherent in the system. Small changes in the product or alternative means for making it can often be simply accomplished, thus allowing easy circumvention of patents and copyrights.

Some nations have been reluctant to extend patent or copyright protection to software, as it represents a new and little-understood legal subject. Others have been quite willing to allow, or even encourage, piracy, which is hard to monitor and prevent.[128] For those countries, there can be no basis for trade in these products. In effect, a new barrier to trade is created, a new form of protectionism. As a result, intellectual-property questions in the late 1980s quite suddenly became important issues in international trade, with the United States in particular strongly advocating the inclusion of property protection in the trade standards of the General Agreement on Tariffs and Trade (GATT).[129]

Many countries, developing ones in particular, have different attitudes toward information than do industrialized countries, arguing that ideas are the property of everyone and cannot and should not be limited. Or, if they accept some limitations in principle, they argue that the technological gap between North and South is so large that industrialized countries should, to compensate for past exploitation, be willing to share some of the new technologies more "equitably," or at least make them available at lower prices.[130]

Lax intellectual-property protection systems, however, can work to the disadvantage of developing countries.[131] Pirating technology limits the transfer of associated technology and condemns countries constantly to have to catch up to work done elsewhere. It also makes it more difficult to build an indigenous technological infrastructure capable of making use of technology that is freely available. And it makes it more difficult for countries to attract foreign r/d investment or to export products to countries with strong property-protection systems.

The arguments over the appropriate level of protection of intellectual property are not new. But as a result of the enhanced importance of technology in national economies and the expanding international diffusion and transfer of technology, the subject is one more that has become of considerably greater importance on the international trade agenda. Nations are seeking to find acceptable ways to rationalize differing patent and copyright policies, with the goal of achieving more nearly common approaches toward intellectual property that can meet the needs of nations in widely varying economic circumstances. It is a goal that will not be easily realized. Thus, in one more way does the advance of science and technology alter (and complicate) traditional areas of international relations.

Command Economies

The sudden collapse of the Communist governments of Eastern Europe and the Soviet Union in 1989 and 1991 had many causes, the poor performance of their economies being among the more important.[132] A major aspect of that poor performance, perhaps the most decisive, was the inability of their command economies to maintain the technological capacity essential for growth, especially in contrast to the vibrant performance of Western market economies.[133]

The problems that those economies have had with respect to technological performance raise the larger issue of whether market economies are inherently better able to develop technological capacity than others. Are there elements inherent in the scientific and technological enterprises, and in what is required for them to flourish, that ensure that some economic systems will perform better than others—and in particular that command economies will inevitably lag behind those that have a basic market structure? One factor alone will not determine economic performance, nor are there pure market or command economies.[134] But the answers to the general question have important consequences for anticipating which economic systems will survive and prosper and which will wither in the future international environment.

It is not difficult to see why it is so hard in a command economy to maintain the conditions for innovation that are so fundamental a characteristic of a well-functioning market economy. Nathan Rosenberg and L. E. Birdzell, Jr., argue, as we discussed in chapter 2, that the essential ingredient for stimulating innovation is an environment that encourages change; in their view, the key development that created the necessary environment in the West was the marriage of science and technology with industry, which resulted in "improved recognition of the possibilities of change, reduced the risks of attempting change, and increased the probable rewards of change."[135]

That environment is precisely what the command economies of Eastern Europe and the Soviet Union did not provide. The incentives were almost reversed. Innovation must by definition produce change, but that carries grave risks: disruption of complex, interlocked production and distribution plans, dependence on unprogrammed supplies that are probably unavailable, uncertain markets for new or improved products, underfulfillment of planning targets, and distortion of structural and personnel relations. The pricing mechanism, determined by factors other than competition in the market, provides no automatic rewards for innovation; in fact, cost reductions or altered input requirements brought about by innovation may cause severe price or supply dislocations.

The vertical command structure, with information flowing up the chain and orders flowing down, is necessary for planning and control, but it has the effect of inhibiting the horizontal communication of ideas and information that is an essential stimulus to innovation. The separation of r/d from production that typically results from the vertical command structure inhibits either the incentive for r/d aimed at improving processes or products, or the setting necessary to make such improvements and to translate them to production. Central planning and control also encourages a preference for autarky in order to isolate the economy from foreign interactions that complicate planning.[136] As a result, competition from abroad is excluded or greatly buffered, further reducing the incentive for innovation that comes from competition or from new ideas.

Thus, the incentives in the system work to penalize rather than encourage innovation. The problem is exacerbated by the demanding production environment of high technology, which requires unprecedented levels of quality control. Without adequate incentives and experience, command economies encounter difficulties in achieving or maintaining adequate production quality that have become major barriers to progress.[137] It is not a surprise that the Soviet Union was unable to compete successfully with the West in high technology.

Import of advanced technologies from market economies is not an alternative means of creating a technologically viable economy, for successful transfer requires recipients with equivalent skills and an environment that can adapt and build on the imported knowledge. Technologies purchased or otherwise acquired from the West were, of course, put to use and— particularly in the high-priority security sector—were successfully applied to military purposes. But those technologies did not become the seed corn for the future that would have kept the technological base abreast of the West's, for high technology has to be at least partly self-sustaining. It can be reinvigorated from external sources, but not wholly sustained by them.

Moreover, complete dependence on foreign technologies that are already embodied in products guarantees a permanent lag in technological capability. Technologies available in the market or already deployed in systems are always obsolete compared to what is under development in the laboratory or is about to be put into production. This consequence of technological dependence, in fact, led to concern on the part of some in the Soviet Union that general dependence on Western technology would not be effective in closing the gap and should not be encouraged; but there was little alternative.[138] There was also concern in the Soviet Union about the direct political effects that would result from the use of technologies that had become essential in the West for competitive economic performance. Effective use of computers, for example, requires that they be widely available and connected to interactive networks at the working level. That implies a willingness to accept unsupervised horizontal communication linkages—anathema to authoritarian societies, as discussed earlier in this chapter—and also opens the door to independent ties to international communications networks. The unwillingness of Soviet authorities to accept that degree of interaction was, in fact, one of the reasons for the absence of a creative computer culture in the Soviet Union.[139] Similarly, the restrictions that prevented unsupervised access to photocopy machines were also barriers to the unfettered communication of ideas and information so essential to economic progress.

Of course, there have been substantial accomplishments in science and technology in the East, and particularly in the Soviet Union; it would be a mistake to paint a completely negative picture. Those nations have been able to produce impressive technologies, especially in space and military hardware, through massive commitment of resources and talent in selected areas, and often through a kind of brute-force approach to technological design. Science in the former Soviet Union was strong in many fields, though not as strong and innovative relative to science in the West as native talents and investments in science should imply.[140]

In general, however, the performance of the scientific and technological elements of the economies of the Soviet Union and Eastern Europe has been at best inefficient and sluggish, with only occasional bright spots—in stark contrast to the vitality and rate of advance in the West.

Recognizing the problem and its importance in coping with the economic deterioration in the Soviet Union, President Gorbachev set improved technological performance as one of his priority goals in his plans for perestroika.[141] In fact, appreciation of the economic significance of competitive technology, the gap in technology between the Soviet Union and the West, and the inability to keep that gap from widening were particularly influential reasons for Gorbachev's conviction that major reform of the system was necessary.[142]

Major reform, however, has implications for the entire ideological underpinning of a regime, not only for the economy. Restructuring, if it is to succeed, must create new levels of autonomy for organizations, reduce hierarchical control of the center, reform the price structure, encourage independent entrepreneurial activities, and lead to other measures that in effect would dismantle the key features of a centrally planned economy. And if the environment for innovation is to be created, there has to be genuine liberalization in the possibilities for permitted horizontal communication among individuals and organizations.

Gorbachev appeared to understand the requirements and implications of perestroika, and in addition realized that he had to enlist the intelligentsia in order to encourage entrepreneurial initiative and participation.[143] The campaigns for glasnost (for greater openness in all aspects of Soviet political and economic affairs) and for *demokratizatsiia* (for greater democracy in the selection of the leadership at all levels of society) were instituted to bring about these essential political and psychological changes.

For all of Gorbachev's appreciation of the nature of the problem, and his success in precipitating massive change in the Soviet Union, he seemed in the end to be unwilling to take the restructuring steps that would undermine the dominant role of the Communist party. His reluctance to move decisively eventually led to the challenge from the conservatives in the party, who attempted to reverse the reforms Gorbachev had introduced.[144] Their failed coup d'état in August 1991 resulted, by the end of the year, in the effective destruction of the power of the Communist party (though not necessarily of former party officials, who typically remained in bureaucratic positions) and the breakup of the Soviet Union into fifteen independent republics, eleven of which have formed a loose Commonwealth of Independent States. Boris Yeltsin, president of the Russian Republic, has been more forceful and committed than was Gorbachev in carrying forward radical economic restructuring in that republic; to varying degrees, the other republics have proceeded along parallel lines.[145]

In early 1992, it is not at all clear what the outcome will be. The economic situation has proven to be much worse than earlier thought; inflation is rampant and economic dislocation severe. There remains strong resistance to change on the part of those who will lose by change, as well as more-general resentment in the public about the sacrifice of long-cherished benefits such as permanent employment, fixed prices, and subsidies for many necessities of daily life. In addition, the relaxation in the bounds of acceptable political behavior and the accompanying easing of restrictions on information have allowed the reemergence of ethnic and national movements that will certainly threaten the future stability of the independent republics and the Commonwealth of Independent States.[146]

It is difficult to be optimistic about the near-term economic prospects for all of the republics, but a return to a centralized Communist state seems highly unlikely. As Michael Mandelbaum observes, "The fragile democratic structures in Russia and the other successor states to the Soviet Union are certainly threatened, but more by chaos and misery than by fanatical ideologies."[147] The nature and scale of assistance from the West will play a role in the economic recovery, particularly in the larger republics of Russia and Ukraine, though it can be only a marginal role at best. Ironically, some of the technologies that the West strove to deny the Soviet Union during the cold war, such as computers and advanced communications technology, are the very ones that will be needed in the monumental task of modernizing the economies of the former Communist states.

China faces similar problems as it also struggles to improve the performance of its command economy and of its science and technology. China elected to scale back, though not abandon, its ambitious restructuring plans (which had been in advance of those of the Soviet Union in tending toward liberalization of the regime) after the events in Tiananmen Square in 1989, and to return to managed information flow. The political costs of liberalization were too high for the regime to accept. Those can be decisions for a limited period of time only if China is ever to be in a position to meet fully the economic aspirations of its people and to fulfill its global ambitions. Presumably, it will take new leadership in China, as it did in the Soviet Union, to bring about the structural changes required for a modern technological economy.

In sum, the characteristics of a command economy are basically inconsistent with achieving the technological performance necessary for economic competitiveness in today's environment. Those economies are based on a fixed and circumscribed menu of goods and services and on a well-defined, stable division of labor among production organizations, for which technological change is a disruptive, disabling intrusion. Command economies cannot compete with, or stay abreast of, those for which continuing technological change is a fundamental element in their operation.

In effect, the demands of international competition in high technology have come to play an important role in determining the economic and, as a result, the political organization of states, a quite significant evolution among the elements of international affairs. Here we see an example in which the characteristics required for the productive pursuit of science and technology are not neutral in their societal effects, but favor the evolution of national economies, and thus of the international system, in one direction rather than another. It is a bias that has tantalizing implications for the likelihood of the ideological convergence of industrial powers. The "end of ideology" is not a new idea, but we can observe here how the requirements of technology contribute to its emergence.[148]

North-South Transfer of Technology and Dependency

The effects of technological change are clearly relevant to the fortunes of third-world as well as industrialized nations and to the relations between them; the discussions of issues such as diffusion of weapons technology, the role of information technologies, intellectual-property rights, and the consequences of different economic systems for technological competence have made this point. Other issues to be considered later in this and in the next chapter, such as the environment, agriculture, and population growth, also involve developing countries directly and have become important aspects of the many and complex relations between the countries of North and South.[149]

Our purposes in this study have led us away from approaching these issues in the broad context of development, in favor of considering them in their own right. It is important, however, to focus separately on one set of issues—those raised by technology transfer between developing and industrialized countries—because of their relevance to the economic aspirations of developing countries and because they offer important examples of the influence of technology on the evolution of the North-South dimension of international relations.

Science and technology enter into the economic issues associated with third-world development in a great variety of ways, the most important being their relationship to economic growth. The difficulty of going beyond a general statement of economic importance, however, is the same as that encountered before. Even if "science-based technology" can be identified as "the major source of economic growth,"[150] spelling out in detail what that should mean for resource allocations has imperfect theoretical and practical underpinning. Changes in many other parameters of a society in addition to science and technology are necessary to realize their benefits, but these parameters prove to be exceedingly difficult to disentangle for policy-planning purposes. Moreover, there is enormous variation in third-world economies and in their governance structures—ranging from the very poor countries, mostly in Africa, to those now called the newly industrializing countries (NICs), mainly in East Asia and moving closer to developed-country status. The role and significance of science and technology are clearly not the same for all, but they are a significant element in all.

Nevertheless, there is widespread appreciation of the general importance of technology—much less so of science—to development, a reflection of its evident role in the economic success of industrialized countries and in the growing proportion of technology-based goods and services in international trade. Sometimes, perhaps too often, technology has been elevated to near-mythic status, as the "magic bullet" that will guarantee development.

Accordingly, technology in particular has come to be an important element in economic planning and investment in developing countries and, as a result, an economically and politically significant aspect of relations between industrial and developing countries. In the process, it has also become a source of many disputes in issues such as trade, employment, energy, patents, weapons proliferation, and environmental degradation, as developing countries seek to advance their interests and as industrial countries adjust to new trade and competitive challenges.[151] Developing countries are often particularly concerned because of their perception that the technological gap between them and the North is widening rather than closing. Technology-based issues, whether in the context of economic assistance or of conflicts of interest, are certain to become a larger part of the relations between industrialized and developing countries as time goes on.

Important as the growth of technology's role in these relations may be, there is no reason to believe this will be other than a gradual evolutionary change, punctuated by the "graduation" of a larger number of countries to industrial status—unless planetary-scale threats, discussed in the next chapter, alter drastically the requirements for North-South cooperation and transfer of resources.

Technology Transfer and Indigenous Capacity

One of the elements that affect the ability of developing countries to assimilate technology effectively, and one that will therefore be a factor in North-South relations long into the future, is the issue encountered frequently in other international contexts—the seemingly simple goal of transferring technological knowledge from one person or institution or country to another. In reality, whether the move is from laboratory to production within a single firm, from market to command economies, or from industrial to developing countries, it is quite difficult to transfer technology effectively if the purpose is more than providing a piece of hardware for use as is.[152] What is there in the transfer and adoption of technology that makes this task so difficult, and how does that difficulty affect North-South relations?

There are many routes by which technology moves within nations or across borders, varying from the actual transfer of a piece of hardware to the education of students. In between is a host of other paths: scientific conferences, publications, patent applications, sales brochures, contract proposals, press briefings, legislative hearings, employee mobility, licenses, acquisition of companies, and others, including some that are outside the law. On occasion, just the demonstration of the feasibility of a technology is sufficient, without any other information, to lead to the knowledge of how to reproduce it or to make use of it.[153] In fact, as noted earlier in other issue areas, it is possible only to delay, not to prevent, the movement of knowledge.

But transfer of technological knowledge is not by itself all that is required for successful technology transfer. Transfers of hardware, such as radios or automobiles, may be relatively straightforward, and the technologies can usually be used directly (at least until they need repair). But if the goal is to transfer technology so that it can be reproduced or can be used as a base for further technological adaptation and development, the requirements are much more demanding. For those purposes, it is essential that the receiver of the technology have the competence to understand it, to carry out r/d that may be required by the particular application or objective, to apply relevant knowledge from other sources, and to add the experience and know-how that can never be fully seen in a final product or a technical description; and the receiver must have the general knowledge required to adapt and to reproduce the technology in its specific setting.[154] In short, the recipient of the technology must have competence not far removed from that of the provider if the transfer is to be successful. That applies whether the transfer is within a firm, between companies, or across borders.

The lesson for developing countries is clear: indigenous capacity in science and technology (for science plays a part in a strong technological capability) are an essential requirement for embedding technology successfully in an economy. Without such capabilities, the transfer of technology from industrial countries will never become an intrinsic element of economic growth.

The need for indigenous capability is probably the strongest lesson that developing countries have learned in the many years of technical-assistance efforts and in the operation of multinational companies. At that level of generality, the lesson gives little guidance with respect to the scale of resources needed or how to allocate them when provided.[155] But it does underline the importance of building resources within a nation, rather than assuming that continued reliance on others is an acceptable alternative.

This conclusion demonstrates one consistent characteristic of the scientific and technological enterprises: they require continued commitment, training and cumulative experience to be successfully pursued. That is, it is not possible for a country or a company—or a university, for that matter—to become accomplished in science and technology simply by means of massive purchase or donation of technology. The building process is slow and demanding, and it requires investment over many years; the importation of technology will provide only a surface veneer until the underlying competence exists to accept the technology and to use it as seed corn for growth.

The extensive U.S. bilateral technical-assistance programs begun by President Truman in 1949 made this mistaken assumption that transfer of technological know-how would be relatively simple, and in addition that technology was all that was required to lead to economic development.[156] Realization of the inadequacy of that approach led, after the 1950s, to much more varied foreign-assistance programs conducted by the United States and other industrialized nations.[157] Those programs continue as important elements of North-South relations, though they usually raise political controversies over quality, size and focus—controversies that will surely grow as global environmental and other issues become more pressing.

Ironically, bilateral programs for technological aid and cooperation, even if of questionable economic effectiveness, have also proven to be politically useful for governments in their international relationships. The evident relevance of technology to national economic interests, coupled with the presumed apolitical character of technology, has made such programs valuable as low-key instruments of foreign policy.[158]

Technology-based Dependencies

International dependencies are an inherent part of an interdependent world, a natural consequence of the integration made possible and stimulated by technological advances. Earlier in this chapter, we explored some of the particular technological forms of dependencies among industrial countries. These dependencies have some similarities with, and some differences from, dependencies between developing and industrial countries. There is disagreement, however, often on ideological grounds, as to whether these latter dependencies are a natural consequence of interdependence or a result of conscious colonial and postcolonial exploitation.[159] We will not attempt to engage that debate; but it is important for our purposes to see in what ways North-South dependencies are sensitive to technological change and to see whether appreciable alteration in their substance or character can be seen or would be expected. Three changes of broad, generic impact stand out; they all are closely related, and all reflect new patterns introduced by science and technology.

ACCESS TO TECHNOLOGY

The first change in the character of dependency relations is a result simply of the greater significance of technology transfer in North-South relations. Most countries of the third world are still dependent on the importation of technology from industrialized countries—some simply for the hardware they are unable to produce at home, others for the continuing flow of knowledge needed to infuse their indigenous scientific and technological efforts.

Part of the rationale for technical-assistance programs for developing countries has been the reduction of the level of this dependency by improving indigenous ability to absorb and build on transferred technology.

The NICs, in particular, have been successful in strengthening their technological capability—which partially explains why they have become NICs—and some, notably Korea and Taiwan, have become substantial producers and exporters of high technology. Few developing countries, however, have yet been able to turn their technological dependence into the more nearly equal mutual dependency characteristic of the relations among industrialized countries.

Dependence on external sources of technology is likely to continue as a fact of life in North-South relations, with little overall change and little that most developing countries can do to reduce the dependency in the short run. Some have tried, with extensive import-substitution policies for technology coupled with protectionist measures to shield infant high-technology industry from external competition. That course of action may reduce dependence for a time; but, as is usual for protectionist measures, it serves to insulate indigenous development from the competitive pressure so important for innovation. The policies may ultimately result in a viable domestic industry, but its products are likely to be higher-cost and not competitive in performance with foreign technologies. Brazil, which followed the protectionist path for its fledgling computer industry, found the policy to be damaging; it has now taken steps to open its economy once again to importation of computers.[160]

COMPARATIVE ADVANTAGE

A second generic change in the nature of dependency relations stems from the effects of technology on comparative advantage in trade and industrial relationships. Until recently, developing countries have been able to capitalize on some aspects of their otherwise-disadvantaged state of development. Low-cost labor, for example, has been an important inducement for attracting manufacturing; a relatively unregulated environment has encouraged the siting of facilities unacceptable for environmental or safety reasons in industrialized nations; and raw materials have long provided essential income for countries whose principal economic asset is a primary commodity.

But, as discussed earlier in this chapter, technological developments in automation and management procedures are reducing the wage content in the cost of many manufactured products and enhancing the value of proximity to suppliers, thus reducing the importance of labor cost in siting decisions. Technology provides means for reducing the local environmental externalities of production processes, thereby challenging the cost advantage of less-regulated foreign sites. And natural resources are becoming a smaller proportion of the value added in manufacturing as technological change allows substitution or design that avoids costly inputs, and as economies move to knowledge-based, high-technology industry with relatively much lower raw-material content in final products.[161]

Thus, the flexibility and options that continuing technological development offers for government and industrial policy erode the comparative advantages of developing countries. In an important sense, the economies of those countries are hostage to technological developments emerging from a system over which they have no control or even input. All economies are to some extent hostage to outcomes of the scientific and technological systems of other countries, systems over which they have only limited control. But the economies of developing countries are likely to be very much more vulnerable, without alternatives or the capacity to design around technological change unfavorable to them.

GLOBAL SYSTEMS

The third generic change in the character of dependency relations is a result of the development of global technological systems for information and other services. These systems—including, for example, space- and ground-based weather, communications, remote-sensing, and navigation systems—are characteristic of the advent of the space age in the late 1950s and are becoming of steadily greater economic importance. Private and public investment in a nation is required both to build the infrastructure necessary to tie in to these systems and to exploit the new services they make possible. Over time, the systems and their many offshoots become integral parts of an economy. Reliance on them grows, with equitable access and assurance of continued availability representing essential elements of system performance.

This system dependence makes the management and operating arrangements an important issue for all participating nations. When a system is owned and operated by an international entity—for example, Intelsat (the organization responsible for operating the primary international satellite communications system)—the structure and competence of the organization must be able to guarantee the access and continuity that participating countries require.[162]

If an economically important system is owned and operated by one nation, there is an implied (it may be explicit) continuing responsibility for providing access and continuity that goes beyond typical bilateral or regional dependencies. An economic function, important to many nations, has been created that is controlled by domestic decision processes in the nation owning and operating the system. Those domestic processes may reflect only weakly, if at all, the international responsibilities the management of the system entails. National budget deficits or domestic inter-agency disputes may become the primary determinants of the fate of the system, rather than the interests of participating nations.

The United States demonstrated how a global system can be vulnerable to domestic considerations in its handling of the Landsat remote-sensing satellite system. After the United States had pioneered the technology, devoted many years to leadership in its use, and actively encouraged investment and participation by many countries, domestic considerations led to reduced government funding and a mismanaged attempt to turn the system over to a private company. There is now a possibility the United States will abandon the technology altogether.[163] If that happens, countries that had invested in ground stations and computer systems for acquiring and using the data will be left without continuing inputs and will eventually have to shift, with considerable disruption, to the systems of other countries. Those other systems, in fact, gained their competitive position in large part as a result of the uncertain U.S. commitment. It is an embarrassing (and unfortunately not unique) episode for the United States, with potentially serious repercussions if the United States gains a reputation as an unreliable guarantor of systems on which other countries become dependent.[164]

These dependencies are not different in kind than the obligations any country takes on when it contracts to provide a particular service or commodity to others. The differences in practice are that these global systems represent new capabilities that did not exist before; are deployed in an international environment, which creates a presumption of international responsibility; and tend to be large and costly and therefore not easily replicated.

Those differences are real and important, and they do represent new patterns of international dependency that stem directly from the outcomes of the technological enterprise. But the dependencies are perhaps not as persistent as might be assumed, in good part because of the nature of other outcomes of the technological enterprise. Continued technological development makes the original technologies less unique: competitors eager to tap the same market are mobilized, alternative technologies to accomplish the same purpose are developed, and lower-cost designs emerge. For example, France has been able to master the technologies necessary to deploy the SPOT resource satellite system; communication cables using optical-fiber technologies have challenged the cost and capacity advantage of communications satellites; and lower-cost launch vehicles reduce the cost of placing systems in orbit.[165]

In fact, the continued technological development that makes substitution for an essential technology easier raises its own problems of depen-

dence. Developing nations may commit themselves to an expensive technological system that subsequently becomes obsolete; they then are
dependent on the sponsoring nation's being willing to continue to provide
service and spare parts for an older, less efficient technology, because the
capital cost of replacement with a better system may be prohibitive.

Thus, these dependencies on global technologies are important additions to existing forms of dependency. The directions of technological
change tend over time to reduce their significance by leading to options
and alternatives, though developing countries may see this as small conso-
lation when they continue to be dependent on a very small number of
systems, and hence nations, if they wish to take advantage of the technologies. One consequence is that many developing countries prefer to see
global technologies managed by international organizations that give
them at least some voice in decision making and a better chance for avoiding dependence on obsolete technologies. There are examples of such
internationally owned systems—Intelsat is the most prominent—but by
and large the industrial countries that first develop and deploy the systems prefer to reap the benefits and to continue their control, rather than
submit to international management.

Old and New Dimensions: Geopolitical Measures

In chapter 3, we traced the evolution of the traditional geopolitical measures of power and influence in the national-security sector as a result of
technological change. Many attributes outside the military/security domain are also relevant to a nation's international position and particularly
to its economic standing; those typically listed in international relations
texts include, in various formulations, geographical position, population
size and characteristics, natural-resource endowments (including energy
fuels), agricultural potential, political stability, and effectiveness of government.[166] We have seen that scientific and technological competence
must now be added; many would include other subjects, such as a nation's
space capability and environmental situation. These are complex factors,
not easily calibrated for their role in determining a nation's international
position. It is important, nevertheless, to examine those we have not yet
considered to understand how these old and new elements of international politics are affected by technological change, and what influence
they have on the evolution of international affairs. We will explore seven
such attributes that are particularly affected by progress in science and
technology: natural resources, population, food and agriculture, energy
(focusing primarily on nuclear power), space, environment, and science
and technology themselves.

Natural Resources

Throughout history, few issues in international affairs have been more prominent in theory and policy than natural resources. From the Trojan War to Japan's attack on the United States in 1941, to more-recent concerns over the Middle East and southern Africa, access to natural resources has been a major factor in foreign policy and has been considered one of the basic elements of geopolitics.[167] Notwithstanding its role in the past, the issue of resource dependency has in recent years been sharply transformed by the new capabilities science and technology offer.

Technological change has long been an important aspect of the issue, gradually altering over time the relative importance of specific resources as new products and processes became strategically and economically important. But its largest recent impact is in its effect on the supply side as it alters in major ways the significance of a nation's natural-resource endowment, generally reducing the importance of resources as a geopolitical factor (with the important exception of oil).

Perhaps the most prevalent confusion over the significance of natural resources to a nation's economic welfare has been about the relationship between supply and price. The confusion stems from the seemingly commonsense view that there is a finite stock of resources on the globe that will eventually be depleted as population and economic growth create ever-larger demands.

Resources are not limited or fixed in size. Their availability is basically a function of their price, which in turn is a measure of their scarcity in relation to demand. A rising price calls forth means of increasing supply or reducing demand: utilizing lower-quality ores, improving the efficiency of use, or finding substitutes.[168] Technological change has long been a key to these processes. Nathan Rosenberg states it succinctly: "[Classical models] drastically underestimated the extent to which technological change could offset, bypass or provide substitutes for increasingly scarce natural resources."[169] Today, the growing technological ability to tailor the design of products or processes for specific purposes or to bypass constraints makes the role of technology even more central to the issue.[170]

Moreover, the evolution of economies toward information- and knowledge-based high-technology industry has had the effect of making sophisticated industrial production much less dependent on raw-material inputs.[171] Semiconductor chips, plastic used in automobiles, and fiberglass cable all require less raw material in relation to cost and output than major industrial products in the past.

These effects of technological change will become more prominent in the future. Society will increasingly be organized around information technologies utilizing relatively fewer natural resources, and the flexibility that is one of the natural outcomes of the scientific and technological enterprises will present increasing options for minimizing undesirable geopolitical resource dependencies.[172] This is, then, another example of a situation in which the outcomes of science and technology tend to favor the evolution of an issue in one direction and not others—in this case leading to reducing the significance of natural-resource dependency as an element of international affairs.

Natural resources as an international issue cannot be totally dismissed, however. It cannot be guaranteed that science and technology will produce a suitable response to a given need at the time the political or economic environment requires it, and there may be many situations in which the development and diffusion of new technologies will be on a longer time scale than the threat to which a response is desired (oil dependence is today the obvious illustration). Moreover, some important and extremely rare materials concentrated in a small number of countries, such as the platinum group, might not be readily replaceable. Finally, the price signals that help to drive the r/d system may not adequately reflect the true scarcity of a resource (such as the environment) or the real price of security of supply, so that developments to affect the use of the resource may not be forthcoming when needed.

Most important, large-volume resource dependencies—fossil fuels in general and oil in particular are the best and perhaps the only major examples—cannot be quickly altered and will therefore remain an important resource issue well into the future. Technology will eventually be able to reduce the importance of oil in international commerce, but the difficulty of developing competitive substitutes and alternatives, and the time necessary to replace massive energy systems, guarantee long-term dependence on that resource.

Though resource dependencies will remain on national foreign-policy agendas, there will be a gradual reduction in the significance of natural-resource endowments (other than oil) as a factor in geopolitics. The resource dependencies that remain will be essentially temporary; the forces at work will lead to, though they cannot guarantee, a reduction of vulnerability to disruption. Some dependencies will become steadily less important through the natural progress of technology; others can be managed through parallel economic and policy measures, such as stockpiling or mandated boosts in prices (through taxes, for example), which will stimulate more rapid development of alternatives and reduce demand.

Thus, though there are many detailed aspects to the role of natural resources, it is clear that, overall, technological change has had a profound effect on the evolution in the meaning and significance of this traditional element of geopolitics. Over many centuries, technological change has altered the relative importance of particular resources and thus the economic and strategic status of countries endowed with those resources. In more recent periods, technological change has made it possible for nations to reduce their dependence on any specific resource and thus to reduce their vulnerability. Continued advances in science and technology will extend this pattern more generally and will gradually remove the issue from the global geopolitical canvas; the only exception is dependence on fossil fuels, which will continue long into the future.

Population

In the past, the significance of population as a factor in international politics stemmed primarily from the relevance of population size both to the size of the military forces that a nation could mobilize and to the power of its economic resource base. As we have seen, those implications have been minimized or entirely altered. Military power has come to depend more on forces-in-being and technological capability than on the number of soldiers that can be put into the field over a long period of time. Population size is not completely irrelevant, however—especially for conflict outside the industrialized countries, as was demonstrated in the long war between Iraq and Iran. Iran's much larger population base allowed that country to resist Iraq's superior armaments for several years, until exhaustion finally set in.[173] But the difference in population between Israel and its Arab neighbors, on the other hand, has so far been neutralized by Israel's far-superior military forces and by its ability to mobilize and field forces rapidly.

The relevance of population size to economic strength has also been rendered less important, or even reversed in meaning, as national economies have come to depend on many other factors. Here, too, however, population size is not totally irrelevant. For Western industrialized countries, a larger population does mean a larger internal market, and it can mean a deeper and broader capacity for innovation. For Japan and the United States, large populations help to maintain their international economic power; part of the motivation for the European Common Market was to create a single economy with market size comparable to or larger than that of the United States. The large population of the Soviet Union, however, did not help it maintain economic competitiveness; other factors were more significant and led to its collapse. For most developing countries, the size and particularly the growth rate of populations have become a serious drain on their economies, with repercussions for the entire international community.

The growth of population, though no longer a primary geopolitical measure, nevertheless is (or should be) a major concern for all nations because of its role in present and future global problems: large-scale migration, pressure on food supplies and land, widespread poverty, and climate change. It may well emerge as the single most serious problem facing the international community and as a component of many others; many believe it already should have that status.[174]

Some 95 percent of the projected increase to over six billion people by the year 2000 will take place in developing countries, guaranteeing continued dominance of population growth in those countries in the next century.[175] On the African continent, the average growth rates is 2.9 percent, the highest of all regions; some countries in East Africa have a rate above 3.5 percent, implying a doubling time of less than twenty years.[176] These are rates much larger than those experienced by European countries at the time of their maximum rate of growth. In China, which has been making a concerted effort to bring its population growth under control, the government was embarrassed in 1989 to find that, even with a supposedly stringent one-child-per-family policy, the total population in that year exceeded 1.1 billion (the target for 1990 was overshot by a full 10 percent in 1988).[177] Projections in 1990 forecast a world population of between ten and fourteen billion by the middle of the twenty-first century, though in 1991 the UN Population Fund recorded, for the first time, a decline in birthrates in all regions.[178]

Advances in science and technology played important roles in the explosive growth of the world's population. They contributed to the economic growth that made a larger population supportable; and they provided the knowledge base that allowed understanding of the role of public-health measures in the reduction of mortality and that made modern sanitation and medical care possible.[179] Today, advances in medical technology and in the ability to produce adequate supplies of food continue to fuel that growth, though there are many other factors that determine fertility rates.

Technology also has an important role to play in reducing population growth—for example, through new methods of birth control that are inexpensive and easy to use in third-world countries. It is very clear, however, that it is not just the availability of technology that is relevant, as important as that is. Many other factors, such as the status of women, the survival rate of children, the state of the economy, religious mores, and cultural and traditional practices are at least as, or more, important. In societies with falling or stable birthrates, several of these factors have evolved sufficiently that, along with the availability of the means (through technology) to avoid conception or to abort pregnancy, they have brought about what has been termed a demographic transition, leading to lower fertility rates through natural choice of the populace.[180]

Two other, quite different aspects of the role of science and technology in population issues deserve to be emphasized. One is the increased importance to an economy of the technological and related skills of its population. The rise of technological competitiveness as a central feature of a global economy has put a premium on the skills, especially scientific and technological skills, of the work force.

Some observers put this factor near the center of their analyses of the future of the international political system. James Rosenau cites the change in the required skills of individuals in the modern, technological world as one of his three key parameters of change.[181] Harold Sprout has argued that population size will be the eventual determinant of geopolitical status, on the grounds that technological skills are the primary requirement for economic and military strength and that the capacity to acquire those skills is evenly distributed in the world's population. The nation with the most people, it follows, will be the strongest.[182]

Sprout's argument is clearly oversimple, ignoring the many factors that go into economic strength. Whether technological skills are as central to the future of the international political system as Rosenau judges them to be, there is no doubt of their importance to national economic performance. The concern over the inadequacy of primary and secondary education in the United States, especially in the sciences, reflects recognition of the significance of the relationship and of the potential international costs of that inadequacy.[183] American universities may generally be judged the best in the world, but the need for higher skills at all levels of advanced technological economies implies that international economic rewards may ultimately go to the nations able to educate the highest proportion of their citizens for modern needs.

The other aspect of the role of science and technology in population issues relates to the age distribution within a nation's population. The extension of life spans made possible by advances in medical science and public health, in tandem with the low fertility rates that have become the norm in industrial countries, confronts those countries with the prospect of a larger proportion of elderly and, in some cases, with actual reductions of total population. By 2010, the proportion of people over sixty-five in Japan is expected to have doubled from its 1980 level—from 9 percent to over 18 percent.[184] Before the reunification of Germany, the total population in the Federal Republic of Germany was expected to *decline* by as much as a third in one generation.[185] The infusion of population into united Germany will certainly prevent such a decline for a time, but the situation is emblematic of what is happening in industrial countries in general. These changes will have an impact on the economic strength of industrial countries as social services consume an increasing share of resources and young workers bear a larger portion of the total economic burden. The changes will also lead to pressures for increased in-migration from developing countries in order to fill essential jobs, likely adding to the ethnic social problems increasingly in evidence in many countries.[186]

It is also worthy of note that in many industrialized nations, including those with the greatest economic power, technological change is contributing to internal demographic shifts that can have important international implications for the future. The responses to a changing economy and to available technologies for birth control have not been uniform among different social groups within nations; the result has been changes in the balances among ethnic, national, and religious populations. Over time, such shifts will alter the nature of many current conflicts—for example, in Israel and South Africa. The effects on the larger industrial countries may also be profound: African-American and Hispanic populations in the United States are growing faster than the non-Hispanic Caucasian population; among the republics of the former Soviet Union, explosive consequences can be expected from the very much higher growth rates of Muslim populations in the Central Asian republics than of the European stock elsewhere.[187] The implications for the countries concerned and for their international relationships are bound to be substantial.

In sum, science and technology have served to reduce the significance of population size among the measures of a nation's geopolitical status, while also playing important roles in the explosive growth of the world's population—growth that is raising many new and dangerous issues for international affairs. Science and technology will be critical components in the efforts to limit that population growth—directly, through the development of new technology, and indirectly, by contributing to the economic growth necessary to bring about demographic transitions within developing nations, where most of the growth is occurring. In technologically more advanced countries, the skills of the population in science and technology, rather than population size, are becoming an important factor in economic strength, and thus a new and significant element in international relationships.

Food and Agriculture

One of the more dramatic doomsday predictions was that of Thomas Malthus in his celebrated study *An Essay on the Principle of Population,* published in 1798. Malthus argued that there is an insurmountable barrier to population expansion because population naturally increases to the limits of subsistence, while food supplies are necessarily constrained by the finite amount of land available. Diminishing returns from increased investment in the scarce resource of land would eventually mean a decline in agricultural output per capita and thus the cessation of population growth.

Obviously, it hasn't worked out that way. Notwithstanding the increase in the world's population from well under one billion in 1800 to five billion by the late 1980s, the per capita production of grain in 1987 was higher than ever before in human history.[188] There are many reasons that food supplies were able to keep up with population growth, the most important (until the end of the nineteenth century) being the economic forces that gave higher returns to farmers for more production and the addition of new lands, particularly in America, devoted to agriculture. Since then, much of the credit for avoiding the predicted diminishing return to effort goes to the new scientific and technological knowledge of crops and farming practices that made greatly increased productivity possible. By the beginning of the twentieth century, the application of science and technology to agriculture had become a formally organized activity, most notably with the establishment of the agricultural research and extension system in the United States.[189]

The most recent dramatic example of the role of science and technology in agriculture came after World War II in what has since been called the "green revolution"—the dedicated commitment of research and field testing for the improvement of seed strains of wheat, corn, rice, and other cereals, particularly for application in developing countries.[190] The results were stunning. Cereal production between 1950 and 1985 outstripped population growth, increasing from roughly 700 million tons to 1,800 million (1.8 billion) tons, an annual growth rate of approximately 2.7 percent. Annual population growth over the same period was roughly 2.0 percent.[191] From 1950 to 1987, the productivity of land more than doubled, from 1.1 to 2.4 tons per hectare.[192] As the World Commission on Environment and Development noted, "Most of it [growth in food production] is due to a phenomenal rise in productivity . . . achieved by: using new seed varieties, . . . applying more chemical fertilizers, . . . [and] using more pesticides and similar chemicals."[193] The new seeds and techniques, a product of targeted r/d, made it possible to realize crop increases that more than matched the growth of population. It is not an exaggeration to say that the green revolution was the essential development over this thirty-five-year period that averted widespread famine and that has kept hunger from becoming a central item on the agenda of international politics.[194]

The technological advances that were responsible for the green revolution came largely from internationally organized and managed laboratories physically established in the developing countries themselves. The first laboratories were founded by American private foundations; later, governments and international organizations followed the successful early models in creating new laboratories. Thirteen of them are supported by a consortium of donor governments, international agencies and philanthropic foundations in the Consultative Group on International Agricultural Research (CGIAR), housed in the Food and Agriculture Organization (FAO).[195]

The green revolution may have been essential to avert famine, but there were costs as well. As with all technological innovation, unplanned side effects accompanied the use of the new technologies. The most evident arose from the requirement for intensive use of fertilizers and pesticides and from the much-increased need for water. The result was increased stress on the local environment, exacerbated water shortages, and, unless adequate credit was made available, inequitable advantages for rich farmers who could afford the larger input requirements.[196]

Now, concerns have been raised that the gains of the green revolution may have run their course, while population continues to climb. In both 1987 and 1988, world grain production fell sharply, following the dramatic rises since 1950.[197] World carryover grain stocks—the amount left unused at the time of the next harvest—fell from 101 days of global supply in early 1987 to 54 days two years later.[198] The better growing conditions and higher prices of 1989 led to a rise in production; but increased consumption exceeded the rise, so that reserves were further depleted.[199] Lester R. Brown, president of the Worldwatch Institute, believes that physical constraints may now mean that there will be diminishing returns to further investments in technology intended to improve productivity.[200]

Adding to this anxiety about the future are the droughts in the United States and other grain-producing countries that accompanied the exceptionally warm years in the 1980s. The concern is that these could be harbingers of the global warming (discussed in the next chapter) that is expected from the greenhouse effect, or at least an indication of what would happen if warming occurred. Reduced water availability in the center of continental land masses and more frequent drought, both predicted consequences of global warming, could depress agricultural productivity throughout the world, as they did for grains in the particularly hot summer of 1988 and for corn several times during the 1980s.[201]

These problems of externalities—increased environmental stress, reduced availability of fresh water, and erosion and exhaustion of land— may, in tandem with continued high rates of population growth, signal a new situation in which agricultural production is inadequate to the need on a global basis.

The scare of the late 1980s appears to have been premature, however: in 1990–91, the world cereal crop was expected to increase by 5 percent over the previous year, with world grain stocks increasing for the first time in three years (though the stock-to-use ratio would still be lower than at the time of the 1988 drought).[202] Short-term fluctuations can be expected, and they tell little; long-term possibilities, however, are very large. It is entirely plausible that realistic economic policies could, for example, improve the incentives for more efficient use of water, recovery of land, and better farming practices.

Most important, it is likely that science and technology will once again offer a way to ameliorate shortages. For example, hydroponic, closed-cycle manufacturing of food that will not require extensive land commitments might be possible within fifty years. And the technological possibilities of the burgeoning field of molecular biology hold the promise of wholly new approaches to the design of crops to meet altered growing conditions and environmental circumstances. That biotechnological revolution has already begun; it seems only a matter of time until its promises will be realized.[203]

As promising as new technologies for food production appear to be, however, technological futures cannot be guaranteed; agricultural applications of genetic technology have proven to be particularly difficult and not yet notably superior to traditional experimental techniques. It is tempting to accept technological optimism and be confident of the timely positive outcomes of science and technology, but they cannot be assured.

There is also a problem of a different kind: the danger that the extensive use of a very limited variety of seed strains—a tendency when favorable new varieties are developed—could make crops on a global basis vulnerable to sudden failure through attack by disease or pests to which those strains are particularly susceptible.[204] Genetic diversity provides protection against such widespread, catastrophic failure; history is replete with cautionary examples, such as the blight that caused the loss of Irish potato crops and the resulting famine in the 1840s, and the phylloxera that destroyed the grapevines of Europe at the end of the nineteenth century (the vines were saved by crossbreeding with American vines that were resistant to the disease).[205] Genetic uniformity in large systems on which many people depend can be a source of dangerous vulnerabilities, an issue to which we will return.[206]

Overall, it is evident that scientific and technological change has been a key factor in making it possible to produce adequate supplies of food, that most fundamental of human needs. Without that capability, world hunger—and probably widespread famine—would almost certainly have been a more pressing international issue than it has been. (The Soviet Union, with its large agricultural potential, also had to rely on imports because of political and structural problems rather than technological shortcomings.)

If the world's agricultural system proves unable to meet the challenges posed by future population growth and environmental change, the issue of food supplies could move sharply to the foreground of international politics. The ability to cope with those challenges is not by any means wholly dependent on progress in science and technology, but satisfactory outcomes from science and technology are a necessary, if not a sufficient, condition. The gradual narrowing of the genetic base of the world's agricultural system may be creating new vulnerabilities that could prove to be important global threats. Science and technology are, as usual, both a central cause of the emergence of the threat and an essential requirement for dealing with it.

Energy: Nuclear Power

The production of energy is the essential means of multiplying the physical power of human beings and of operating the technological core of human society. It is the indispensible element of the economies of all nations, with corresponding significance in international affairs and in any geopolitical accounting. It is so central to national economic security and strength that all countries must perforce be concerned about the uninterrupted availability of energy at economically acceptable costs. Science and technology have for centuries been integral both to the provision and use of energy and to the gradual reduction of costs and increase in flexibility of use that have made possible the rapid rise in living standards, particularly since the start of the industrial revolution.

Though science and technology are at the very heart of the subject, their relevance in detail to any given energy-policy issue is highly variable—dominant for some issues, inconsequential for others. Science and technology are germane to both the supply and demand sides of the energy equation; new and improved technologies alter the availability and cost of energy and of the resources needed to produce it, and other technologies affect the requirements for energy and the efficiency of use.

Energy is so pervasive an aspect of modern society that substantial portions of this entire study could have been organized around one or another of its facets. In fact, many of the important relationships of science and technology to energy are encompassed in other subjects discussed earlier—such as dependency relationships, natural resources, transfer of technology, and international trade—or will be examined in the next chapter, when we discuss the global environmental effects of energy emissions.

Rather than attempt a complete discussion of technology-related energy issues—which would repeat many points already made and whose length would distort the entire study—we will examine the introduction of nuclear power as an informative surrogate for all issues. It is relevant to a substantial number of more general energy issues, raises several of the problems that will be inherent in the development and introduction of any new energy technologies, and presents important new issues in international affairs on its own. Moreover, it was a sufficiently dramatic innovation when introduced that potentially large economic and political changes on the international scene were expected to flow from it. It is important for our purposes to understand why that did not happen and what lessons that unexpected outcome may have for other technologies that will be candidates for implementation in the future.

Before we turn to nuclear power, however, a few general observations about the role of science and technology in energy can be made. First, the scale of the world's energy systems, as for all large technology-intensive systems, precludes rapid systems-wide transitions from one form of energy technology to another. Whether the goal is substitution of fuel, or the installation of more-efficient technology for production of electricity, or implementation of alternative means of producing power, long lead times are necessarily involved.[207] The same is true for the r/d to develop alternative energy-production technologies. Investments must be made with expectations that payoffs will come only after many years, which in most cases will mean some sort of government support of r/d. Even when the r/d lead time is not long, as for efficiency improvements of some end-use technologies (e.g., more-efficient light bulbs or electric generators), the time it takes for the technology to diffuse and come into general use will be materially affected by the size of the system that has to be penetrated.

A second observation is partially a consequence of the first: existing international oil dependencies cannot be appreciably modified in the near term by technological developments. Even if there were acceptable technological alternatives in prospect, there is no realistic substitute that could quickly replace the uniquely large volume of oil that moves in international trade.

Third, even in a subject as technology-intensive as energy, nontechnical aspects are typically at least as significant as the technological aspects. National policies on taxes, subsidies, efficiency regulations, and similar matters are critical not only in determining the use of technology, but also in influencing the nature of the technology that is developed. Similarly, the international environment, both political and physical, will have much to say about the nature of the technology that emerges.

Finally, notwithstanding the relevance of nontechnical matters, it is ultimately the outcome of energy-related r/d that will make possible eventual transitions in energy technologies on both the supply and demand sides. Whether it be some form of renewable energy source—fission or fusion power, solar energy, or synfuels; or new capabilities for conservation, improved efficiency, and perhaps alternative mobile energy-production systems; or, more likely, a combination of those and other possibilities—it will be the results of r/d that provide the basic choices for society. That r/d, it is worth noting, will be heavily motivated by foreign policy objectives—reducing dependence and minimizing global environmental effects in particular—so that the progress and direction of science and technology will themselves be a function of the pressures stemming from international affairs.

There will not be any shortcuts in this process. Some technologies may be more promising and more easily implemented than others, but r/d is not going to turn up surprises that will have immediate, revolutionary effects. Even if the supposed discovery of cold fusion had been real, the time unavoidably required for turning a discovery into a genuine innovation, the inevitable questions about cost, safety, and side effects, and the scale of the systems to be replaced would have dictated a long time between the discovery and its widespread application.[208] Though the possibility of surprise always exists in science, the likelihood of the sudden emergence of a completely revolutionary concept and its immediate embodiment in a practical technology is hopelessly remote.

With these observations in mind, let us turn to the example of nuclear power.

The use of the atom's energy as a power source was not, of course, the primary goal of the r/d that provided the knowledge base for nuclear power. The driving objective was the development of an atomic bomb, first recognized as a realistic possibility after the 1938 Hahn-Strassmann experiment (discussed in chapter 3). There had been earlier speculations about the possibility of tapping the energy of the atom, reflected in novels of H. G. Wells and Harold Nicolson; and, in fact, the necessary scientific evidence was in hand earlier—but not correctly interpreted.[209] An atomic bomb, however, remained only fanciful speculation until 1938, when, as news of the experiment rapidly spread, scientists all over the world immediately realized what might be possible. It was a dramatic demonstration that basic research, thought to be without useful application, could in fact have enormous practical significance.

After the initial discovery, it took seven years and what was for then a massive commitment of resources ($2 billion) and personnel to build a bomb in the United States and use it against Japan. The possibility of tapping the atom's energy for peaceful purposes was evident from the beginning; the development of a self-sustaining nuclear "pile," first achieved at the University of Chicago in 1942 in the course of the A-bomb project, demonstrated that it could be done.[210] Nevertheless, it took fifteen years and considerable further r/d to realize a reactor suitable for the production of electricity for commercial purposes.[211] In fact, much of the progress was still a by-product of military objectives; the first direct application of reactors was for submarine-propulsion systems, and the reactors developed for that purpose were the basis of the early designs for commercial use.[212]

Several objectives motivated the development of the civil applications of the technology. It offered the promise of an inexhaustible source of energy for the future, thus relieving any concerns about the extent of fossil-fuel supplies or the danger of dependency on other nations for energy resources. Moreover, it was assumed to be almost cost-free and thus able to provide, at very low cost, the energy essential for all economies. In a now-ironic phrase that continues to haunt the industry, electricity from atomic energy was predicted to be "too cheap to meter."[213]

It also appeared that atomic energy would be the basis for a large and important industry that would benefit whichever country established early leadership. The assumption that there would be an economic bonanza led to considerable friction between the United States and the United Kingdom during the war, a difficulty settled only in the 1943 Hyde Park agreement between Roosevelt and Churchill committing the United States to share commercially relevant information about the technology.[214] The friction was renewed after the war when the U.S. Atomic Energy Act of 1946 reneged on that commitment and restricted the release of atomic-energy information to any other country (a decision reversed, however, in the 1954 revision of the act).[215]

Development of commercial applications was also seen as necessary to justify the unprecedented expenditure of public funds for the Manhattan Project. More important, peaceful use was increasingly seen as a counterbalance to the morally repugnant military applications—especially after the development of hydrogen weapons, which were several orders of magnitude more powerful than the Hiroshima bomb. The latter was one of the motivations for President Eisenhower's Atoms for Peace program, launched in 1953, which was the beginning of the conscious use of peaceful applications of atomic technology in support of foreign policy.[216]

For the United States, which had plentiful fossil-fuel reserves, dependence on supplies from abroad was not a particularly serious issue in the 1940s and 1950s. But other nations were highly dependent on external sources of fuel. Japan (which has to import almost all of its fossil fuels), France, and Germany saw nuclear power as a way of reducing that dependence, which had figured as a major cause of war in the Pacific and was an important factor in the conduct of World War II in Europe.[217] This new technology provided the promise of a plentiful domestic resource that could be substituted for a highly undesirable external dependency. Parallel views emerged in the United States when the oil shocks of the 1970s dramatically raised the price of oil and vividly demonstrated how dependent on external sources of fuel the United States had become.

Finally, growing concern over the damaging local environmental effects of fossil-fuel emissions emphasized the potential attractiveness of nuclear power as a "clean" fuel.

Thus, nuclear power, in a remarkably brief period, emerged from r/d as a technology that promised to be a substitute for increasingly scarce and geopolitically poorly located resources, that would provide new, low-cost sources of the energy essential for continued economic growth, that would avoid the negative externalities of fossil-fuel combustion, and that would itself be the basis of a potentially very large and profitable industry. When the political benefits of providing a counterweight to military applications and of being useful as an instrument of foreign policy are added, it is clear that the promises of this dramatic, new, quintessentially twentieth-century technology were substantial indeed, with major potential implications for international economic and political affairs.

What went wrong? In the early 1990s, the nuclear-power industry is healthy only in Japan and France, and to some extent in Germany. Seventy-five percent of the electricity in France is generated in nuclear-power plants, 27 percent in Japan, and about 13 percent in the republics of the former Soviet Union.[218] There are operating plants in a large number of countries—more than a hundred in the United States, accounting for some 20 percent of electricity generation in that country.[219] But new orders are few; in the United States, there have been none since 1974. The reasons for the decay of the promise of nuclear power are many and complex, but some of the key elements are particularly instructive from our perspective.[220]

The first was the false promise of a free lunch, or in this case its equivalent in a pure "technological fix." The "costless" energy of the atom had to be translated into a functioning, efficient, safe technology; that process resulted, as it always must, in a technology that was no longer costless. Whether the price of the resulting energy would be competitive or not depended on the price of alternatives. Until the Organization of Petroleum Exporting Countries (OPEC) forced a price rise in the 1970s, the cost of oil remained low, making even the unrealistically low predictions of nuclear-energy prices only marginally of interest. After the oil shocks and their accompanying price rises, the projected costs of nuclear power appeared to be low enough that they led to a spurt of plant orders and construction.

Those early cost assumptions for nuclear power in fact proved to be considerably overoptimistic; there were unexpected technical difficulties, and the problems of design, construction, licensing, safety, and waste disposal proved to be much larger than had been expected. Coupled with the gradual fall in the real price of fossil fuels during the 1980s and the falloff of electricity demand after the original steep price rises, these problems defused what incentive there was in most countries for building new plants. Japan and France are still expanding their nuclear-power commitments, presumably because security of supply is more important to them than whatever price differentials and other difficulties there may be.

Cost was by no means the only problem. Safety concerns became a major impediment as the accidents at Three Mile Island and Chernobyl sensitized the public to the dangers inherent in the technology. The continuing issue of low-level radiation around power plants is controversial and emotional. And the coupling in the public's mind of nuclear power with nuclear weapons and with the unseen danger of radioactivity led to fierce resistance in many countries.[221] France and Japan have been able to avoid decisive public opposition, at least so far, in part because of their more centralized and elitist governmental systems and in part because the advantages of the technology have been widely accepted as being of overriding importance.[222]

Both of those nations have also managed their programs much more effectively than has the United States, keeping costs lower and inspiring confidence in the safety of the installations.[223] Nevertheless, both nation's programs—in fact, all nuclear-power programs—are in an important sense hostage to the weakest programs. It is a different form of dependency, but it is dependency nonetheless. Nuclear accidents wherever they occur will tend to undermine confidence even in well-run programs. The safety of the poorly designed plants in Eastern Europe and the Soviet Union are a source of great concern; the fates of reactors in Western countries may well hang on whether there is another accident on the scale of Chernobyl.[224] Nations that have become heavily dependent on nuclear power, notably France and Japan, may yet rue their substitution of nuclear dependence for dependence on foreign sources of fossil fuel.

Disposal of nuclear waste has also turned out to be a major political problem of international proportions. Disposal at sea of any kind of nuclear waste is no longer allowed by international treaty, and most nations are having difficulty finding politically acceptable disposal sites even for low-level waste on land.[225] No nation has yet been able to select a site for high-level waste, which would contain highly toxic plutonium with a half-life of twenty-four thousand years, as well as other extremely dangerous transuranic waste products.[226] For now, such waste is stored primarily on-site at nuclear power plants, not a suitable solution for the problem of long-term removal of the waste from the environment.

The extremely long times involved in storing and isolating nuclear waste appear to raise a quite new dimension in the otherwise familiar public-policy issue of the estimation and management of risk. Because of the time scale, there is no way of ever demonstrating conclusively that a disposal method will not sooner or later have undesirable effects on the environment or be vulnerable to unauthorized retrieval. All issues of risk in a society raise questions of assessment in the face of uncertainty. But society would be vulnerable to the hazards associated with nuclear waste for a dramatically long time—the wastes would require management for longer than human society has been organized. This prospect, coupled with the sensitivity to anything nuclear, makes the nuclear-waste issue one of the most intractable that must be dealt with in any political process.

In fact, however, some non-nuclear hazardous wastes pose toxicity problems at least as serious, and such wastes may *never* decay. The long time involved in radioactive decay is thus something of a red herring; it does not make the problem of isolation and continued monitoring of nuclear waste necessarily harder than for other waste streams. Society has not shown as much concern for seriously hazardous wastes as it has for nuclear-related waste, however, a fact that emphasizes how the nuclear element alters the perception of the issue.

There is another critical problem with the technology that has important implications for the international community: the danger of proliferation of nuclear weapons. The argument is that widespread use of nuclear power would make fissionable material that is suitable for application to weapons commonly available, thus lowering the barriers to proliferation. As discussed in chapter 3, this view has led to strong diplomatic efforts to contain the risk, through negotiation of the Non-Proliferation Treaty (NPT), establishment of the International Atomic Energy Agency (IAEA) as a monitoring body to deter diversion of fissionable material, and attempts to dissuade countries from "closing the fuel cycle," in order to keep plutonium out of normal commerce.[227]

The IAEA deserves particular mention, for it is an example of an international response to the political implications of a new technology and has proven to be one of the more effective of the specialized agencies associated with the UN system. The creation of the organization, originally proposed by the United States, reflected the growing concern in the 1950s about the possibilities and dangers of proliferation. The earlier proposal in the Baruch Plan for the UN to control all nuclear materials had failed because of Soviet objections.[228] The IAEA necessarily reflected more-modest goals, but the United States wanted to slow what appeared at that time to be the rapid spread of atomic-bomb programs (the Soviet Union had provided substantial nuclear assistance to the Chinese) and to create a framework for international control. The Soviet Union was openly hostile at first, but gradually changed its attitude, and eventually became an enthusiastic supporter. In fact, the United States and the USSR in effect became joint guarantors of the agency's management and effective operation.[229]

The agency is based in essence on a bargain between the nuclear powers and the developing countries. The nuclear powers make the technology available for peaceful purposes through technical assistance; in return, non-nuclear countries agree to a regime of international inspection to deter any diversion of fissionable material to military purposes. The IAEA was responsible for the inspection, or "safeguards," system; and it later also became the action agency for carrying out the safeguards provisions of the NPT.[230]

The safeguards program is particularly interesting, for it involves more significant intrusion into the internal affairs of nations by a formal international inspectorate established for that purpose than is found in any other subject (with the possible exception of oversight of financial matters by the World Bank and International Monetary Fund).[231] The inspections do not extend to the nuclear powers (though the United States and the former USSR were willing to allow demonstration inspections), nor do they cover all of the nuclear programs of other signatories—as the case of Iraq has demonstrated. The IAEA experience may be important, however, as a precedent for international oversight of national programs. It is a rare example of willingness to sacrifice national autonomy to achieve a larger, common purpose.

In the early, heady days of nuclear-power optimism, the United States, and then to a lesser extent other countries, sought to use their knowledge of the technology as an instrument of foreign policy. There were important political purposes to be served: enhancing bilateral relations, demonstrating technological leadership, and establishing controlled access to nuclear power in order to lessen the danger of proliferation. The Atoms for Peace program, which distributed research reactors to many third-world countries, represented one of the more ambitious attempts by the United States to use its technological leadership in a concerted way to serve such foreign-policy purposes.[232]

The results of the program cannot be definitively assessed, though in the 1950s and early 1960s it undoubtedly enhanced the image of U.S. technological strength, an image that after 1957 was being challenged in the public mind in many countries by the space spectaculars of the Soviet Union. It may also have added to the danger of proliferation (though the intent was the opposite) by accelerating the spread of knowledge about atomic energy. That knowledge would have inexorably spread in any case, and the existence of the program did make it easier to develop international agreements and mechanisms of control. But it is not certain that the benefits outweighed the costs.[233]

In sum, the development and application of the new technology of nuclear power has introduced several new and potentially significant elements into international politics. It did not, however, overturn the economics and technology of energy, as was originally expected, and it thus has had a relatively marginal influence on energy politics. A small number of nations are now less dependent on external fuel supplies than they were before, but they may also now be hostage to the safe performance of nuclear installations in other countries, performance over which they have no control. An ephemeral and uncertain dependence has been substituted for a clear and certain one.

The new elements that have been introduced into international affairs relate specifically to the special characteristics of nuclear technology, in particular its relation to the spread of weapons and to the environmental and safety risks associated with its use. The first concern has stimulated extensive international activity and has spawned, *inter alia*, an international organization to which sovereign nations (albeit excluding the major nuclear powers) have ceded more-intrusive rights than they have to any other organization. That development may not be enough to prevent the eventual diffusion of nuclear weapons; but if the political climate for international cooperation continues to improve, it may be a measure of what is possible when the threat of proliferation becomes more immediate.

The environmental and safety risks have also led to considerable international activity, and much more will undoubtedly be forthcoming, especially with regard to global effects. Safety concerns will continue to hound the industry; any new accidents—and there are sure to be some—will cause political waves far from their country of origin. New nuclear technologies presently being studied that entail fewer risks and externalities may be more readily accepted in the future, but the present subjective resistance to anything nuclear will apply to them as well.[234] The threat of major global warming largely as a result of the burning of fossil fuels, a threat we discuss in chapter 5, could change the public's attitude toward nuclear power; but by the time nuclear-power plants could be installed in sufficient numbers, other, less dangerous alternative non-nuclear technologies may also be relevant. Unprecedented policy issues have been brought to public consciousness as the long time scale for the decay of nuclear waste forces consideration today of matters with consequences extending thousands of years in the future; so far, political processes have not proven able to cope with this dimension of the challenge.

Nuclear power has other lessons. Several aspects of its history are characteristic of the introduction of large-scale technologies more generally, such as the delay and cost that inevitably accompany the reduction to practice of new knowledge. Not only are the final costs likely to be higher than first estimated, but the competitive threat to existing technologies is likely to lead to developments in those technologies that will in turn lower their costs, making it harder for the new technology to compete. In addition, the apparent economic rewards of creating a dominant new industry will not necessarily go to the first entrant, and may be ephemeral in any case.

Similarly, externalities will emerge that were not anticipated and usually could not have been predicted. In the case of nuclear power, the U.S. agency responsible for promoting development made the problem of externalities worse, innocently or otherwise, by tending to downplay problems of safety or waste disposal so as not to interfere with atomic-weapons development and rapid deployment of peaceful nuclear power.[235]

Thus, nuclear-power technology has not had the profound influence in the economics and politics of energy that many had expected; it has, however, resulted in important elements of change in international affairs, though not fundamental shifts. The outcome may be different if, or when, a crisis of some kind occurs, whether it be one state attacking another with nuclear weapons, or one or more serious nuclear-reactor accidents. Until that time, however, it can be said that power produced from the energy of the atom has proven to be only an important, not a revolutionary, new factor in the evolution of international affairs.

Space

Space is an entirely new arena of state action, one that has raised a panoply of issues between and among nations since the dawn of the space age in 1957. Like nuclear energy, this subject would not appear on the international agenda but for progress in science and technology in this century.

In the early days of space endeavor, the performance of the superpowers in space came to be interpreted as a measure of their relative economic, military, and scientific strength. In public diplomacy, the Soviet Union assiduously capitalized on its space leadership with the message that its political system had made it possible to overtake and surpass the United States in general scientific and technological accomplishment.[236] The message was made even more credible by the depth of surprise and shock in the United States over the Soviet launch of the first earth satellite, Sputnik, and by the apparent validation of the Soviet claim of strategic-missile superiority. The size of the Soviet payloads inferentially demonstrated a rocket capability sufficient for intercontinental missiles.[237] During that early period, space accomplishment became, in effect, a factor in geopolitics as a surrogate for other elements of a nation's power and influence.

The response to the apparent "defeat" in space was a crash mobilization of resources in the United States to meet what was generally accepted as a massive threat from the new capabilities of the Soviet Union. The mobilization included not only a major national commitment to space, but also accelerated weapons development, new programs for science and technology education, and innovations in the policy process that gave the president direct assistance from the scientific and technological community.[238] Over time, it became evident that the inference that the Soviet Union led in weapons and in science and technology was unjustified, and that its accomplishments in space were not, in fact, accurate indicators of its broader capabilities.

Gradually, capability in space technology faded as a geopolitical factor, as economic applications in space became more common and widespread, as military capabilities matured and became roughly balanced, and as space activities came to be seen as having only a limited relationship to economic strength. Competence in space technologies is not irrelevant to a nation's international political position, however. All industrial nations of substantial size, and some developing countries (India and China in particular), feel it necessary to participate in space programs; and an increasing number have found it wise—for economic, prestige, or security reasons—to develop an independent space capability. Major space programs are carried forward by the larger powers, with commitments of resources that could not be justified by the expected short- or medium-term economic returns: the multibillion-dollar manned space station program of the United States, for example; or the manned programs of the former USSR, conducted in the face of a seriously deteriorating economic situation. Apparently, space activities have not lost all relevance to international status, though the persistence of large programs has as much to do with the bureaucratic goals of space agencies as it does with economic, scientific, or prestige goals.[239] In any case, these activities no longer have the deep political significance in their own right that they held on the international scene for a brief period.

In fact, with the exception of scientific research, activities in space are no longer an end in themselves, but a means for accomplishing other national objectives. Space is, for example, a major arena in the security sector, in much the same sense that other environments—land, sea, and air—are relevant to security goals. Ballistic missiles travel through space; surveillance systems are designed for observation from space; C^3I systems are stationed in space; the illusory strategic-missile-defense system, SDI, would be largely space-based; and arms-control measures depend heavily on space technologies for monitoring of compliance with agreements. The security aspects of space are now a permanent element of international politics, their importance a derivative of the importance of security issues more generally.

The direct economic benefits of space applications have steadily gained significance. Technological systems for communications, weather forecasting, resource assessment, navigation, broadcasting, even provision of launch services, have become common and increasingly important to national economies. The expansion of these systems has been one of the factors increasing the degree of interdependence among nations, not only because of the role the systems play, but also because of the international cooperation many require for their design, launch, management, and regulation.

Potential space applications have also been a spur to r/d in frontier high-technology industry, leading to differential economic benefits as some nations prove to be more successful than others in those domains. The total economic return from national investment in space assets, however, remains controversial. Spin-off claims continue to be asserted by space enthusiasts and denied by the critics, especially with respect to manned space efforts.[240] However, the benefits of significant investments in space automation and robotics could be important contributors to other high-technology industries.[241]

Space may be a physically infinite medium, but for some applications there are significant constraints on various parameters that require regulation and resource allocation. Those applications that depend on a satellite's remaining in a fixed position over the earth, for example, require that satellites be stationed in a geosynchronous orbit over the equator at a height of 22,300 miles so that they will move in synchronism with the earth's rotation. The number of positions available for satellites in that orbit is not unlimited, so that slots must be assigned according to agreed rules and performance standards must be set. The International Telecommunication Union (ITU) has been given the responsibility to manage that process, in line with its related responsibility for regulation of the radio-frequency spectrum.[242] It is a typical example of how the international community, desiring to exploit a global technology, must turn to international mechanisms of some kind to manage the necessary regulatory functions.

Other space applications have led to the formation of new international organizations, such as Intelsat for satellite-based communications; all such applications have introduced new elements that require international negotiations and cooperation.

For scientific research, the ability to move outside the earth's protective atmospheric blanket to study the solar system and deep space from new vantage points has, not surprisingly, been enormously stimulating, greatly adding to knowledge and understanding. All space-capable nations have mounted scientific experiments, with research usually touted as the primary, or at least the first-mentioned, justification for their space programs. Space satellites also provide platforms for the study of the near-earth environment, a capability of growing importance because of the mounting concern over global environmental degradation.[243] President

Bush has even urged a Space Exploration Initiative (SEI) to develop a permanent presence on the moon early in the next century and to achieve a manned landing on Mars by 2019.[244] Space spectaculars have obviously not disappeared from the international agenda, though in any given program there are always multiple objectives to be served—not excluding the bureaucratic goal of seeking larger budgets.[245]

All of these activities are inherently international, if for no other reason than that space is itself an international environment, belonging to no nation and, as such, common property. Whether it is the "common heritage" of humanity and thus owned by all, or owned by none and thus open to all, is an important question when it comes to sharing and to compensation for use. That remains an issue on the international agenda.[246] More generally, most of the uses of space require international cooperation or negotiation, so that space-related questions are now common in international affairs. Major projects such as a manned space station, a lunar base, or a manned Mars expedition, for example, will require contributions by more than one nation if they are to proceed.[247]

Thus, space has become a new locale for issues in international affairs, directly as an outgrowth of scientific and technological progress. However, the novelty of its first appearance on the scene has largely worn off, to be replaced by issues that bear close resemblance to familiar subjects. Space-related matters have introduced new issues to world politics and have brought novel considerations and expertise to prominence. For a time, space appeared to be a new and significant factor in geopolitical relationships; but that significance has faded without altering, in any lasting way, deeper aspects of international affairs. The advances in science and technology that made space exploration and technology possible have created a new arena for international politics, but they have not altered the underpinning of those politics in any substantial respect.[248]

Environment

Environmental issues and science and technology are inherently closely related. Technology has had a key role in industrial and economic growth, and it is the negative externalities of that growth that have fueled much of the concern since the 1960s over degradation of the environment. Environmental movements with considerable followings have grown within and across countries; some have had substantial political effects, particularly in Western and Eastern Europe. Green parties have become potent political forces in most Western industrialized countries; and in Eastern Europe, mobilization around environmental issues played a part in the political upheavals in those societies in the late 1980s.[249]

Technology is not the cause of environmental damage; it is the application of technology without adequate knowledge about, attention to, or concern for its externalities that has led to many environmental problems. It is by no means always new technologies that are involved; often, pollution is the result of the deployment of old, polluting technologies even when cleaner ones are available. And considerable environmental change and degradation is simply a result of an expanding world population exploiting its surroundings for fuel, food, and shelter, as people have always done.[250] Technology, nevertheless, has often been seen as the villain.

But technology, and science as well, is also a necessary part of the solution. It is with the essential contribution of science and technology that environmental problems are frequently first realized, their effects analyzed and understood, the policy options laid out, and new technologies developed to avoid dangers or to minimize damage. In fact, technologies that are environmentally friendly or that reduce existing environmental problems may offer large new markets for industries prepared to meet the demand. Japan has taken steps to invest substantial sums in r/d with the intention of leading those new markets.[251]

As a result of the recognition of the seriousness of environmental issues and the more-intense public attention to them, the environment has become a steadily more important aspect of domestic and international politics, and it is likely to become even more important as population and wealth continue to grow.[252] The state of the environment is not normally considered a geopolitical measure of a country's international position, though it can be a significant indicator of the health of an economy and had a geopolitical role in the fall of Communist governments in Eastern Europe. It is, rather, a new category of issues that will appear more often on the international agenda.

As a first approximation, these issues are another manifestation and consequence of global integration, with many of the same implications for international relationships that arise in that context; as such, they are not unfamiliar to international affairs. Joint use of boundary rivers for water supply and waste disposal, cross-border air flows that carry effluents from one country to another, disposal of waste in international waters, and other activities of one nation that affect the environmental conditions of others are now common in the interactions of countries and are often politically and economically difficult subjects to navigate. They are handled through quite-standard activities—negotiation, establishment of standards and regulation, cooperative research, allocation of public goods (and bads), agreements to reduce or phase out dangerous practices, and assignment of responsibilities to new or existing international organizations.

Bilateral and multilateral organizations with environmental responsibilities, such as the Commission for the River Rhine, have existed for more than a century, and new ones are being created.[253] The European Community has established an environmental directorate, with the ambition of aggressively expanding communitywide rules.[254] The appalling environmental conditions in Eastern Europe that came to light with the opening of those societies have put regional pollution issues high on the political agenda, as those countries attempt to rebuild their economies and as Western European nations try to limit cross-border damage.[255] The UN and its specialized agencies have been increasingly involved in global environmental issues since the Stockholm Conference on the Human Environment in 1972, a landmark in the evolution of worldwide environmental consciousness that also led directly to the creation of the UN Environment Program.[256]

And environmental issues have become embroiled in new ways in trade questions. Nations have begun to seek to extend their environmental policies to other nations by limiting imports that were not produced or caught under conditions that meet those domestically determined policies. In 1991, the General Agreement on Tariffs and Trade (GATT) ruled that the United States could not ban the importation of tuna from Mexico on the basis of U.S. legislation that prohibited tuna imports from any country that killed more dolphins than the American tuna fleet.[257] The GATT ruling considered the American law as a nontariff barrier to trade, whether or not the law had other motivations.

The growing international economic and health consequences of environmental issues also illustrate vividly how much international dependencies have increased—particularly how much states have become dependent on the internal practices of others, often having no control or influence over those practices. The situation in Europe is an unfortunate example. Before the economic and political disintegration of the Eastern European countries, the decision making in their command economies gave little or no consideration to environmental consequences. The results not only were disastrous for them, but could and did affect Western Europe as well, with the potential for much more serious damage in the West if no action was taken.[258] Yet the nations of Western Europe had no direct way to intervene in those decisions or to influence them. The poor design and maintenance of many of the nuclear reactors in Eastern Europe threaten Western Europe with Chernobyl-like dangers, a situation that demonstrates all too well the level of cross-border dependencies on the unilateral policies of other states. So, too, does the environmental damage that resulted from the Gulf War, when, in an act of senseless and barbaric revenge, Saddam Hussein set fire to six hundred Kuwaiti oil wells, threatening damage to regions far from the scene of hostilities.[259]

Thus, the importance and prevalence of environmental issues have increased on the international scene, in some cases with substantial implications for underlying relationships and for the development of new international organizations and functions. These issues are another example of the growth of interdependence, with its accompanying constraints on the autonomy of governments to set policy unilaterally within their own borders.

Nevertheless, most environmental issues, are not inherently different in nature from their predecessors on the international agenda. They represent an upgrading of those past issues in that they have greater political importance and more-significant implications for central concepts such as dependency and autonomy; but they represent only gradual, evolutionary change. Their geopolitical significance is, so far, modest, except in the special circumstances of the extreme damage that resulted from indifferent attitudes in the Eastern European command economies.

One class of environmental issue could, however, become very much more important in international affairs, both to the evolution of international affairs and ultimately to the very structure of the international system. These are the issues that are global in their implications and that threaten to damage the ecosystem of the entire planet. Their importance and potential implications for international affairs call for separate analysis, which is offered in chapter 5.

Science and Technology

The relevance of competence in science and technology to almost all significant geopolitical factors implies, as we have noted often before, that the scientific and technological capacity of a nation—its breadth and depth, its innovative spirit, the skills and imagination of its workers, the creativity of its managers, the quality of the ties between laboratory and production, the support and interest of the public and of their political leaders, the resources devoted to it, the strength of the educational system—is now a major geopolitical factor in its own right that must be considered along with those of a more traditional character.

As with the others, the relative strength of the scientific and technological systems in different countries is not fully measurable, for it consists of many qualitative and attitudinal variables as well as quantitative ones. The great advantage the United States held in science and technology after World War II has decreased—not necessarily because of U.S. decline, but because the capacity of other nations has relatively improved, as it had to do. The gap between the United States and the Soviet Union before that country broke apart, however, steadily widened, in part because of the Soviet Union's economic difficulties but also because its economic structure proved over time to be incompatible with the requirements of a high-technology economy.

In the light of all the previous discussions, little more need be said here to demonstrate the geopolitical importance of a nation's scientific and technological capacity. However, it is well to recognize that, as in so many areas of assessment of a nation's strengths and weaknesses, perception is at times as important as reality, especially when reality may not be quantifiable or testable. Over time, however, the true quality of a nation's science and technology will affect its accomplishments in all the derivative areas that in practice determine its international position. In particular, the critical role of science and technology in national economic performance will likely continue to be a prime geopolitical factor in the coming decades, and thus a major determinant of a nation's international power and influence.

Large Systems

One of the characteristic consequences of the introduction of many technologies is the growth of large systems made necessary or desirable by the features of the technology. Energy, transportation, and communications technologies in particular have this system-forcing characteristic. Even technologies that appear at one level to have opposite effects (the development of the automobile, for example, represented a decentralized alternative to railroad and streetcar systems) are dependent on elaborate infrastructures (road and fuel-distribution networks) that are large systems indeed.

Large systems have, in fact, become ubiquitous in the deployment of technology, and they are becoming increasingly international. Examples are many: transportation systems able to provide large volumes of oil, food, raw materials, and manufactured products anywhere in the world; airlines able to connect far-flung points on tight (usually) timetables, using a common language and common regulations; interconnected railroad networks throughout Europe, with trains that run frequently and at high speeds; large-capacity electricity-distribution systems within nations and, increasingly, across borders; global-scale space applications that tie together national facilities, such as systems for communications, navigation, weather forecasting, resource sensing, and now direct broadcasting; and a telephone network that connects essentially all extant telephones in one global system—certainly the most massive example so far.

In addition to the telephone, other large systems associated with information technologies are the most momentous examples today, with the emergence of a range of information capabilities interacting in massive transnational networks. Corporations, securities markets, governments, scientific laboratories, news organizations, universities, military forces, and individuals are able to interact on networks that essentially ignore borders and geography in their operation. As continued development brings down unit costs and expands capabilities, more use is made of the technologies, and new ways are found to capitalize on the opportunities they provide.

It would be banal to point out the growing economic and social importance of these large systems, several of which have already engaged our attention. Two aspects deserve our special attention, however, because of the new elements they introduce in international affairs, elements that derive from particular characteristics of large systems.

Vulnerability

We have already discussed, early in this chapter, the vulnerability of global financial markets to systemic problems or breakdown. The same vulnerability exists in other industries and activities—such as airlines, electric-power distribution networks, synoptic scientific research, and the telephone itself—in which real-time, uninterrupted access to a vital system component, whether it be information, airports, electricity, or fuel, is essential to the functioning of the system. In effect, whenever various societal activities are tied together in large networks, a dependence is created on the continued and effective functioning of those networks. The more demanding are the requirements for immediate responsiveness in complex systems—that is, the more tightly coupled they are—the more serious will be the consequences of interruption.[260]

This vulnerability is widely recognized; designers of systems incur considerable costs in money and performance to minimize vulnerability by means of design redundancy, security measures, and quality control. Such protection cannot be perfect, however, for there is always a trade-off between increased security against disruption and the efficient operation of the system. If the system is to operate at all, breakdown can never be completely ruled out.

The question of interest for us is whether this systemic vulnerability is a growing danger or whether it is comparable to vulnerabilities of the past. Qualitatively, the answer has to be that risks are growing: global society is more vulnerable as it comes to rely more extensively on large technology-based systems. The conclusion must be cautiously made, for technology also provides the palliative—the means to reduce vulnerability, or at least mitigate its effects. But it seems inevitable that, on balance, vulnerability will become a steadily more dangerous and destabilizing factor in international affairs.

Global information systems, for example, create opportunity for operational intervention that can be highly disruptive, or worse. Inputs can be devised that propagate through the system, "infecting" individual computers, erasing data, or otherwise changing the operating characteristics of the system.[261] Dubbed "computer viruses" because they function with many similarities to biological viruses, these programs are a constant threat to information networks. Their ability to surmount or avoid protective measures in the system, is dependent on the competence of the virus designer. One person or a few can do the job; a large organizational effort is not required.[262] Technology-based methods and equipment can help to protect against such viruses, but the only foolproof means of protecting a computer—isolating it from the network—would be self-defeating by denying the use of the network.[263] More sensible and less draconian measures of protection necessarily involve trade-offs between safety and efficiency; proof against all risk is not possible if the benefits of the computer network are to be realized.

Intentional disruption is not the only danger. The electricity blackout in the U.S. Northeast in 1965 was caused by the failure of a relay in one distribution point of the electric grid.[264] Weather disturbances at a key airport can cause worldwide air-traffic disruptions. In global financial markets, as we have seen, the failure of one component of a system could bring the whole system down, with large financial consequences throughout the world. A minor software error in a two-million-line computer code controlling long-distance routing for the AT&T telephone system shut down AT&T long-distance lines for nine hours on January 15, 1990, and a blown fuse in Manhattan cut off all long-distance AT&T calls into and out of New York City on September 17, 1991, virtually shutting down airline traffic in the New York area.[265]

Building on the flexibility science and technology can provide, measures can be taken to reduce the danger of all these vulnerabilities. Backup facilities can prevent loss of data or contact in case of interruption of network communications. Improved landing equipment for aircraft and better forecasting can reduce disruptions due to weather. Technological design can improve quality and redundancy in electric grids and provide alternative emergency power supplies. In some cases, it is a matter simply of understanding better through research just what the dangers are and of designing the systems in the knowledge of those dangers. On the other hand, the costs of protective measures, as well as the growing reliance on software that is exceedingly hard to make secure against error, will make truly adequate protection steadily more difficult.

The diffusion of military weapons poses a different, and very serious, threat to vulnerable systems. The spread of weapons with ever-greater performance capabilities—concealability as well as power, reach, and accuracy—makes large systems vulnerable to disruption by rogue governments or terrorists. The importance of large systems to an economy, coupled with their vulnerability, makes them particularly attractive targets.

In addition, the vulnerability of a system is always likely to remain at least one step ahead of measures to reduce that vulnerability. The simple reason is that it is difficult to appreciate the extent of vulnerability until a breakdown occurs. Moreover, calling attention to the need for protection can be seen as implying weakness in new systems, which designers would prefer not to acknowledge. In any case, protective measures add to costs at the time a new system is implemented when there is always an incentive to minimize costs and emphasize benefits. Dangers tend naturally to be underplayed, as a result, until they are vividly demonstrated.

The historian William McNeill even posits a principle of "conservation of catastrophes" suggesting that all advances in civilization necessarily are accompanied by a kind of cumulative risk of catastrophe. He writes, "Perhaps we should recognize that risk of catastrophe is the underside of the human condition—a price we pay for being able to alter natural balances and to transform the face of the earth through collective effort and the use of tools."[266]

Thus, the vulnerability of large systems, though rarely noticed until disruption or catastrophe occurs, has become an element in international relationships and may become a very much larger danger. It represents a new and destabilizing factor in international affairs, not as a daily element in relationships, but as the source of occasional crises that follow failures, from natural or malign causes, of large critical systems. Some of those crises may not cause severe dislocation; others can be very severe indeed, with large economic, human, and political costs.

Initial Standards

Another, important aspect of large technological systems is the significance to national economic competitiveness of the operating parameters and standards set for such systems at the time they are introduced. This significance is not a new concept; industries have long understood the potential economic value of being the first to market a systems technology that forces others to adopt operating standards that favor their product, and governments have long realized the importance of setting standards that give a domestic company an advantage.[267] The French, for example, resisted European adoption of American color-television standards in the 1970s in order to protect their own television industry.[268] Computer manufacturers in the United States compete fiercely over the selection of a common operating system for computers, a choice which directly affects computer compatibility and which could determine the fate of companies and networks.[269] As large technological systems grow to become global in scale, the economic stakes can become much larger, with the initial standards a critical element in deciding the winners and losers.

Two new major developments are on the table in the 1990s, both with large-system implications and correspondingly large economic consequences. The more important for the long run is the proposal to establish an ISDN (integrated services digital network), which, when it materializes, will eventually be the most extensive technological system ever built. An ISDN is envisaged as a single global, high-capacity information network able to encompass all voice, data, and video networks that now constitute separate systems. The idea is simply to make a more flexible, higher-capacity system, to which all users of current systems (and many new users) can join at lower cost and for higher added value.

The idea may be simple, but the choices are complex, their consequences far-reaching, and the controversies enormous. Profound questions are raised, for example, about centralization versus decentralization in system architecture, control of and access to the system, designs that favor some equipment manufacturers and some users over others, what kinds of businesses are intended to be the primary consumers, means for protecting privacy and intellectual property, and the ability of governments to regulate, and even to monitor, telecommunications traffic.[270] The schedule for implementation of the system and for appropriate developmental steps is itself a source of controversy. The outcomes of these debates will affect the comparative advantage of companies and possibly of whole industries, with potentially very large and differential effects on national economies.

The issues are being debated within governments and negotiated in international governmental forums such as the ITU and the International Standards Organization (ISO). Telecommunications, electronics, and computer companies are very much involved, as are government agencies—particularly the PTTs (post, telephone, and telegraph ministries) in countries that still give those organizations responsibility for telecommunications—and multinational corporations and service providers that are heavy users of information networks. Some countries, such as France, have been able to develop reasonably coherent national positions and have even taken early steps in implementation to put themselves in a favorable position.[271] Others are far from settled in their views.

Whatever the outcome, an ISDN is an example of a new level of global-systems development that will have major economic consequences for nations, consequences heavily influenced by early choices in the design of the system. The arcane nature of the issues, however—a result of the esoteric technologies involved—makes it extremely difficult to have effective lay participation in policy debates, notwithstanding their importance to an economy.

Another major system in the early stages of development, high-definition television (HDTV), is an example of a different type, one somewhat closer to past experience. Since the 1970s, the Japanese have been developing a radically improved television broadcast and receiver system, which will provide television images with much higher resolution and thus greater clarity.[272] The Europeans have had a project with similar objectives under the EUREKA program since 1985.[273] Many different approaches to the design of the system are possible; the Japanese version, developed by NHK, the Japanese national television network, was the first to be demonstrated. It would require doubling the number of lines on the television screen, along with other changes, and would not be compatible with current receivers. The European approach would allow compatibility. The United States entered the field belatedly; but technological advances have made an all-digital approach—in which the United States is very strong—likely to be the future direction of television development, thus making the United States competitive in the race.[274] The U.S. Department of Defense has provided limited support, justified on the grounds of possible military applications.[275]

The battle is initially over the market for the next generation of television receivers. The company or companies controlling the technology chosen as the basis for a worldwide standard will have an enormous market advantage. Delays in reaching agreement on one standard, or the adoption of multiple standards, would give lagging companies a chance to catch up. The U.S. Federal Communications Commission (FCC) said in 1988 that an American standard would have to be an over-the-air system, thus limiting the bandwidth available and requiring effective compatibility with existing television sets.[276] That decision negated the Japanese lead and gave U.S. firms breathing space to develop an alternative to the Japanese system. Cooperation among European countries in developing their own system—which from the start was chosen to be different from the Japanese approach—is similarly motivated by the desire to give European companies a chance of succeeding in their own market.[277]

Television is a large market, but the HDTV-standards issue also has broader and possibly more important ramifications. Some see HDTV as embodying a technology that will have not only major effects in the television industry, but also broad application in the computer and telecommunications industries.[278] If so, that technology will have repercussions throughout high-technology industry, with important military, medical, graphics, and many other imaging applications, while at the same time creating a much larger market for semiconductors. If the initial standards favored American companies, it would create an opportunity for a resurgence of the U.S. electronics industry, which has suffered from the near-total domination of the television market by the Japanese. And it would be a significant element in the larger competition in high-technology production and trade, an element that could have more-profound long-run consequences for national economic power and influence.

The formal agreement on HDTV standards, which will be finally settled in negotiations in the ITU, will not determine the final outcome of the race; but it will have a major effect on which countries and companies can make a run for it. Though this decision may not be nearly as important as that on ISDN standards, its magnitude does make it of considerable economic significance to all nations that will be system suppliers or users.

Both of these proposals are symptomatic of the critical and growing importance to national economies of the determination of the operating parameters of large technological systems. These are issues that cannot be relegated to purely "technical" discussions, for they now weigh heavily as a factor in the economic fate of nations.

Some Summary Comments

Extensive evolution in the nature and substance of issues is the evident result of the interaction of science and technology with international politics and economics. Some aspects of that evolution have had significant effects on the underlying elements of the international system; other aspects have led only to incremental and often ambiguous effects, though they may well be of substantial importance to the conduct of international affairs and could lead to considerable change in the future. The primary dimensions of that evolution can be briefly summarized.

The most striking changes, and the ones that come closest to being discontinuities in international affairs, are those that affect the internal economic and political structure of states. Command economies, as we have seen, are unable to provide the environment for technological innovation necessary to be competitive in high technology with market-oriented economies. In effect, a command economy is no longer a viable form of economic organization if a nation wishes to be able to compete in a domain that is now central to economic growth and power. On the political side, the introduction of new technologies, particularly information technologies, has become a critical factor—perhaps *the* critical factor—in subverting the centralization of political power in authoritarian regimes. Though these technologies can also be used to help such regimes maintain power, their consequences on balance promote openness within a society and the decentralization of power.

More generally, the integration of economies, the growth of large technological systems vulnerable to disruption, and the multiplication of nongovernmental institutions and activities, all strongly aided by technological change, have led to growing limitations on governments' autonomy and freedom of action. These limitations, though consequential (especially for the conduct of international relations), are in my view less significant for the independent authority of national governments than rhetoric often proclaims. Rather, the evolution of this aspect of the international system is gradual, with science and technology also providing governments with new and powerful levers for maintaining influence over policy.

Economic competition, and the foreign-trade stage on which the competition is joined, is another major issue that has been substantially affected by science and technology. The increasing economic importance of technology-intensive products in international trade has altered the traditional determinants of trade relationships, making comparative advantage a supple concept, capable of being created and molded by national policies to a degree never seen in the past. And that has served to enlarge the scope of trade negotiations, as the sources of comparative advantage now reach more deeply into a nation's structure, culture, and policies. Important as these changes are, however, they do not appear to be discontinuities from the past, unless the possibility of new critical technological dependencies that would dramatically increase the technological dominance of one nation turns out to be realized. In my view, the evidence for the emergence of such dependencies is weak, at best.

Most other familiar political and economic issues have undergone substantial evolution as the international scene has been altered by technological change. New forms of dependency have emerged, or old ones been neutralized, particularly between the nations of North and South. Questions of access to technology, the terms on which transfer of technology can and should take place, the dependence on international technological systems, and the need for indigenous capacity in science and technology have become key issues in North-South debates. In addition, developing countries have found that their traditional advantage in low-wage manufacturing or raw-material production, for example, is eroding as the policy options technology makes available alter the determinants of their comparative advantage. Among industrial countries, a new form of technology-related dependency has emerged: nations with large nuclear-power programs find their programs to be, in effect, hostage to the continued safe performance of other nations' nuclear-power industries, over which they have little influence.

Conventional geopolitical factors have been modified, replaced, supplemented, or made irrelevant. The significance of both natural-resource endowment and population size has been greatly reduced—in the case of population, probably reversed from its traditional value—as large populations (at least in developing countries) have tended to be a drag on economic growth and welfare. Agricultural potential remains important, but may come to mean less if developments in science and technology make it possible for food production to keep up with population growth without the need for increased commitment of land. Energy-resource endowments of fossil fuels will continue to be significant well into the future—probably more so as industrialization expands—while the high promise of nuclear power has, for now, been largely deflated for all but a handful of countries. Nuclear technology is an entirely new and dangerous subject in world affairs, one that raises the nasty issues of weapons proliferation and storage of extremely hazardous by-products over millennia. The history of its development also demonstrates the excessive optimism often associated with the introduction of new technology. Space, another new subject in world politics introduced as a result of expanding technological capabilities, was at first seen as a major arena for geopolitical competition, but has come to be only another realm for the more-prosaic activities of research and economic exploitation.

Finally, science and technology themselves have become important geopolitical factors, in reflection of their role in all of these and other issues. The agenda of international politics has been enormously broadened because of the results of the scientific and technological enterprises, with science and technology firmly implanted as significant elements of a nation's political and economic welfare.

The effects of all these developments on international affairs have certainly been substantial, but surprisingly little has changed in the fundamental elements of the international system. The changes that have taken place—in the actors, the conditions, and the subjects of international affairs—should be seen as evolutionary only, rather than as constituting change in central organizing principles. This conclusion, as we have noted before, is ultimately a matter of personal judgment that we will discuss more fully in the concluding chapter.

Five

Global Dangers

NATIONS HAVE become familiar with, perhaps even reconciled to, living with the apocalyptic threat of nuclear weapons, the most destructive devices ever made possible by scientific and technological advance. Since 1945, international politics has had to cope with the existence of those weapons and with the knowledge that their use in any quantity would wreak havoc on civilization, and could possibly destroy human life on the planet. The international political system has accommodated to the new situation, with some important changes but without the massive transformations in the behavior of states or in the structure of the system that many believed essential if catastrophe was to be avoided.

As the nuclear threat recedes, at least in its virulent cold-war form, new threats are emerging—threats that are characteristically less immediate, more subtle in their onset, less certain in development, and more dispersed in both their causes and effects. But these threats ultimately also pose dangers to life on the planet, through change in climate, spread of major diseases, vulnerability of biologically specific crop strains, loss of species, deterioration of growing conditions, and other effects, all appropriately encompassed in the notion of global environmental change.

In contrast to nuclear weapons, these threats are typically less closely related to the results of the laboratory; rather, they are primarily the unplanned by-product of growth in human population and wealth: the innumerable incremental externalities that arise from housing, feeding, entertaining, moving, and supplying an increasing number of people, all of whom are striving to increase their standard of living and quality of life. As these externalities aggregate, they lead to impacts on the environment that are global in scale and scope. In this sense, global environmental change grows out of linkages among almost all facets of human affairs—science and technology among them—and its consequences correspondingly affect them all.

Science and technology have, however, an additional and quite critical role with regard to these issues: providing the analytical capability to identify the existence of environmental dangers, to assess the facts, to understand the consequences, and finally to develop the means of coping with them. This is a role that is becoming increasingly important and to which we will pay particular attention later in this chapter.

Predictions of global-scale environmental change due to human activities have been voiced for quite a few years.[1] In the early 1970s, a spate of analyses unfortunately included some with such weak methodological bases and exaggerated apocalyptic conclusions that legitimate concerns were debased.[2] The issues have come into sharper focus since then, however, and the pace of scientific research on many of the elements of environmental change has been substantially increased.[3] Curiously, until the early 1990s there were few substantial political analyses that explored the broader international political consequences of global environmental change, notwithstanding growing international political activity and controversy on issues such as the threatening deterioration of the stratospheric ozone layer and the possible rise of the average temperature of the atmosphere.[4]

Global warming—the warming of the earth's atmosphere as a result of human activities—has become the most politically visibile of the global environmental issues; it provides an exemplary case for exploration in detail because of its many implications of direct interest to our study. Moreover, it is a useful surrogate for most of the other issues that can be considered "global dangers," such as deforestation, ozone depletion, loss of species, desertification, and pandemics. To some degree, it encompasses the others; certainly affects them, and raises questions none of the others adequately illustrate on their own.

In fact, global warming may pose more directly than any other issue since the advent of nuclear weapons the question of whether technological change is leading, or will lead, to a revolutionary transformation in international politics. The time scale of change, the magnitude and irreversibility of effects, the direct costs and opportunity costs of mitigation, the true costs of delay or of adaptation, the number of people affected, the effects on other crucial issues, even the question of human survival—all compounded by great uncertainty—conspire to make the issue a stunning illustration of an unprecedented development in human affairs. The usual meaning of cooperation among states does not convey an adequate picture of the degree of interaction among sovereign nations that would be required to meet the threat if the effects at the upper bounds of some of the scientific forecasts materialize. It is clearly a subject that justifies our close examination.

Global Warming

The basic greenhouse effect that leads to global warming is not a new concept, nor is it in doubt. In fact, it is essential for life on earth. Without the warming action of the atmospheric blanket, the surface would be approximately 35°C cooler and unable to support life as we know it, or possibly any life at all.[5] The earth's atmosphere is largely transparent to the incoming shortwave energy from the sun; some is reflected back into space, but a portion is absorbed by the surface and clouds and reradiated as long-wave (infrared) energy, that is, as radiant heat. Several gases absorb energy at these wavelengths and in turn warm the lower atmosphere. Gases (other than water vapor) contributing most to the greenhouse effect include carbon dioxide (CO_2), methane (CH_4), ground-level ozone, nitrous oxide (N_2O), and the entirely man-made chlorofluorocarbons (CFCs).[6]

The atmospheric concentrations of these greenhouse gases have been increasing rapidly as a product of human activities, on a scale sufficient to raise the possibility of appreciable alteration of climate and weather. The concentration of CO_2, the gas that accounts for approximately half of the warming effect, has increased some 25 percent in a century, largely as a result of the destruction of forests and the emissions from the burning of fossil fuels. The human sources of the other gases are primarily agriculture, mining, motor vehicles, and industry; almost all human activities are involved in adding greenhouse gases to the atmosphere or removing some from the atmosphere.[7]

The increase of CO_2 in the atmosphere can be seen most easily in the now-famous data from the Mauna Loa Observatory in Hawaii (see fig. 1). The short-term fluctuations are seasonal; they correspond to photosynthesis with uptake of CO_2 in the growing season and respiration and oxidation during the winter.

The increase in CO_2 released to the atmosphere correlates with the increase in the burning of fossil fuels for energy production and industrial use. Figure 2 shows the dramatic rise in anthropogenic production of carbon since early in the industrial revolution.

However, only about 40 percent of the carbon released through human activities actually stays in the atmosphere.[8] The location of most of the remainder is still unknown, nor is it known how rapidly it will be taken up by various carbon sinks.[9]

More than six to seven gigatons (billion metric tons) of carbon in excess of natural sources is annually being released to the atmosphere (as of 1989). This is only a small percentage (roughly 3.5 percent) of the total carbon released into the atmosphere each year from respiration of plants and soils and from physiochemical processes at the sea surface.[10] But the roughly three gigatons per year that remains in the atmosphere is enough to threaten climate alteration, and is a consequence entirely of human activities.

Most attention has been given to carbon dioxide, but the concentrations of other gases released by human activities that absorb infrared radiation are also increasing. The gases vary in their absorption characteristics and residence times in the atmosphere, so that their relative contributions are difficult to estimate. A UN scientific assessment panel calculated the warming potential relative to CO_2, over one hundred years, to be 11 times

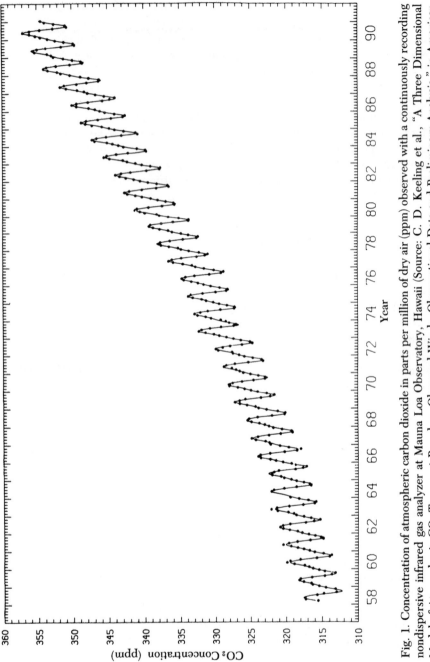

Fig. 1. Concentration of atmospheric carbon dioxide in parts per million of dry air (ppm) observed with a continuously recording nondispersive infrared gas analyzer at Mauna Loa Observatory, Hawaii (Source: C. D. Keeling et al., "A Three Dimensional Model of Atmospheric CO_2 Transport Based on Observed Winds: Observational Data and Preliminary Analysis," in American Geophysical Union, *Aspects of Climate Variability in the Pacific and the Western Americas*, Geophysical Monograph, vol. 55 [Nov. 1989], app. A)

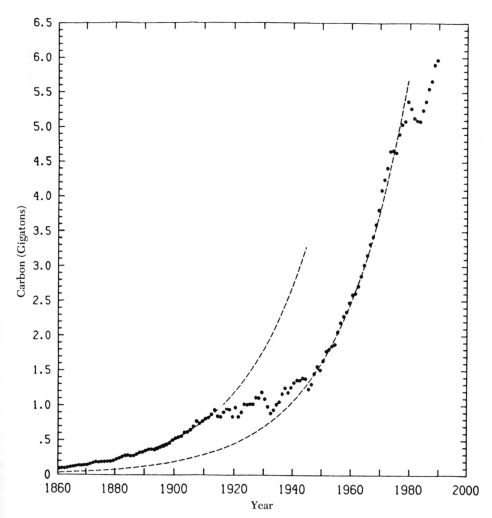

Fig. 2. Annual production of carbon dioxide caused by human industrial activities. Exponential growth, shown by the dashed lines, is twice interrupted—from 1914 to 1945 (a reflection of the two world wars and the worldwide economic depression between these wars) and from 1980 to 1984 (a reflection mainly of the Iran-Iraq war); the period since 1985, although too short to be proven so, appears to exhibit renewed exponential growth (Source: before 1950, C. D. Keeling, "Industrial Production of Carbon Dioxide from Fossil Fuels and Limestone," *Tellus* 25 [1973]: 174–98; from 1950 to 1988, G. Marland, "Global and National CO_2 Emissions from Fossil Fuel Burning, Cement Production, and Gas Flaring," *Trends '90* [Carbon Dioxide Information Analysis Center, Oak Ridge National Laboratory], 1990, p. 93)

for methane, 270 for nitrous oxide, and 3,400 to 7,100 for various kinds of CFCs.[11]

It is obvious that if the concentration of CFCs were allowed to increase substantially, those gases would eventually become the major cause for concern. Fortunately (from this point of view), CFCs that are used widely as aerosols and refrigerants have been shown to be the primary cause of the depletion of stratospheric ozone, a dangerous problem that is an issue largely, though not entirely, distinct from global warming. An international agreement in 1987, the Montreal Protocol, established the commitment for reduction in the use of CFCs; subsequent modification of the protocol in 1990 called for a complete phaseout by the end of the century for industrial countries and ten years later for developing countries.[12] Unfortunately, some of the replacements being considered for the ozone-damaging CFCs are themselves serious contributors to the greenhouse problem; and residence times on the order of centuries for some CFCs already released indicate they will continue to absorb infrared radiation (and destroy ozone) long into the next century.[13]

Methane is a different and very puzzling problem. The concentration of methane is rising even faster than that of CO_2 (1 to 2 percent per year, from a preindustrial value of 0.7 parts per million (ppm) to almost 1.6 ppm in 1989), but not all the sources and sinks are known or well understood.[14] It is released from mining activities, natural-gas-distribution systems, waste dumps, digestive processes of ruminant animals, microbial activity in seasonally flooded agriculture (such as paddy rice), and deforestation. These sources do not account fully for the observed increase, especially since methane measurements are rising much more steeply at the poles than would be expected.[15] It is possible that there is unaccounted release of methane from natural sources in the ground, such as from the warming tundra, or that methane is remaining longer in the atmosphere as a second-order effect of increased combustion of carbon-based fuels.

Methane is the primary constituent of natural gas; since it has one-fifth the carbon content of coal, natural gas is a more attractive fuel than coal from the perspective of the greenhouse effect. It should be noted, however, that even though the residence time of methane is lower than that of carbon dioxide, its roughly twenty times greater effectiveness as an absorber of radiation means that there is no gain to be had from its use if there is appreciable leakage in the supply system for natural gas.[16]

Nitrous oxide is seen to be less of a problem; it is increasing at less than 0.2 percent per year, though it has a residence time of over one hundred years (that of methane is only about ten years). The sources of nitrous oxide are also not thoroughly understood, though it is known to be one of the products of waste disposal and of the breakdown of ammonia used as a fertilizer.[17]

Tropospheric ozone—ozone at the lowest level of the atmosphere—may become more serious. Its concentration is increasing, and it is an effective absorber of infrared radiation. Ozone forms photochemically from urban smog and thus is an indirect product of automobile combustion; but there is inadequate understanding of all of its sources and of the complex chemical processes of its formation.

If their differing residence times and effectiveness as absorbers of radiation are taken into account, these other gases together add up to an effect roughly equal to that of CO_2. Table 4 shows the various gases' relative contributions to total greenhouse-gas emissions in 1985 and their primary sources.[18] The persistence of the gases and the current level of emissions mean that some gases will remain for a long time in the atmosphere even if emissions are not allowed to grow.

This part of the story is straightforward and generally not in dispute. Estimations of the effects, however, are less definitive. They are the result of theoretical analysis, calculation using complex models of the atmosphere and oceans, and interpretation of a limited but growing body of measurements. Important uncertainties and unknowns emerge, and controversies over the models and their interpretation necessarily arise. The more prominent analyses by authoritative scientific bodies—for example, the National Academy of Sciences and the Intergovernmental Panel on Climate Change (IPCC)—coalesce around the estimate that an increase in greenhouse gases to the equivalent of a doubling of the atmospheric concentration of CO_2 would ultimately increase average global temperature by 1°–5°C (less at the equator but up to twice as much at the poles), leading to average surface temperatures higher than any recorded in human history.[19]

When that temperature rise would occur depends on the rate of emission of the gases, the persistence of the emitted gases in the atmosphere, and the rate of transfer of carbon to the oceans and other sinks, as well as on complex feedback effects of clouds and oceans, effects not now understood. Typical analyses predict that, in the absence of substantial restrictions on emissions of greenhouse gases and with no large surges or dips in economic growth, an increase of gas concentrations equivalent to a doubling of CO_2 will occur by the middle of the twenty-first century. Because of time lags in the ecological system, the actual temperature increase at that point might be approximately one-half of the ultimate "equilibrium" temperature increase for that gas concentration.[20] The first report of the IPCC included a consensus prediction of an increase of 1°C by 2025 and 3°C by 2200.[21]

The effects on the earth's surface of such a temperature rise are not predictable in detail, for they would vary regionally, and even locally, and the models cannot now deal with fine structure. In general, however,

TABLE 4
Estimated 1985 Global Greenhouse-Gas Emissions from Human Activities

	Greenhouse-Gas Emissions (Mt/yr)	CO_2-Equivalent Emissions[a] (Mt/yr)	
CO_2 Emissions			
Commercial energy	18,800	18,800	(57)
Tropical deforestation	2,600	2,600	(8)
Other	400	400	(1)
Total	21,800	21,800	(66)
CH_4 Emissions			
Fuel production	60	1,300	(4)
Enteric fermentation	70	1,500	(5)
Rice cultivation	110	2,300	(7)
Landfills	30	600	(2)
Tropical deforestation	20	400	(1)
Other	30	600	(2)
Total	320	6,700	(20)[b]
CFC-11 and CFC-12 Emissions			
Total	0.6	3,200	(10)
N_2O Emissions			
Coal combustion	1	290	(>1)
Fertilizer use	1.5	440	(1)
Gain of cultivated land	0.4	120	(>1)
Tropical deforestation	0.5	150	(>1)
Fuel wood and industrial biomass	0.2	60	(>1)
Agricultural wastes	0.4	120	(>1)
Total	4	1,180	(4)
Total		32,880	(100)

[a] CO_2-equivalent emissions are calculated from the greenhouse-gas emissions column by using the following multipliers:

CO_2	1
CH_4	21
CFC-11 and -12	5,400
N_2O	290

Numbers in parentheses are percentages of total.

[b] Total does not sum due to rounding errors.

Source: Reprinted with permission from *Policy Implications of Greenhouse Warming*, copyright 1991 by the National Academy Press, Washington, D.C. (adapted from U.S. Department of Energy, *The Economics of Long-Term Global Climate Change: A Preliminary Assessment—Report of an Interagency Task Force* [Springfield, Va.: National Technical Information Service, 1990]).

Note: Mt/yr = million (10^6) metric tons (t) per year. All entries are rounded because the exact values are controversial.

rough predictions based on the models indicate reduced availability of fresh water in summer growing seasons in inner areas of the continents with increased precipitation along the coasts, migrations of agricultural growing regions, increased desertification, more frequent and violent storms, and more weather extremes.[22] As a result of the average-temperature increase, there could be a rise in sea level of as much as two feet by the middle of the next century, which would, among other consequences, make low-lying areas and countries already under flood threat, such as Bangladesh, particularly vulnerable to storm surges.[23] If the forecasts are correct, all of this will happen more rapidly than such changes at any time in the last 160,000 years, raising questions of unanticipated and possibly serious transient or nonlinear effects, such as shifts of major ocean currents or slippage of the West Antarctic ice sheet.[24]

Not only are these forecasts uncertain and controversial, they also depend on developments in the economy, future technological change, and policy choices in coming decades. Nor is it clear that all the effects of warming will be undesirable. Humankind has in general preferred warmer climates, and some activities, such as food production in the high latitudes, may be enhanced. Some localities and nations may actually benefit in comparison with others or in absolute terms. The costs of adaptation to climate change, furthermore, are at least as uncertain.

This schematic picture of global warming and its possible physical and socioeconomic effects is frequently presented in more apocalyptic terms; certainly there has been no dearth of analyses, inspired by visions of catastrophe, laying out what *must* be done to reduce CO_2 and other greenhouse-gas emissions and proposing targets for the United States and other countries in order to "stabilize" the atmosphere.[25] Most countries have found it exceedingly difficult to reach agreement on appropriate policies, however, and the internal policy debates are reflected in the international arena as nations attempt to decide how to cope with this global issue. The subject is rich in complexity and detail, with several unusual characteristics that make it considerably more daunting than other global issues and particularly relevant to our study.

Special Characteristics of Global Warming

INTERDEPENDENCE OF INTERESTS

The first special characteristic is that global warming, or more generally global climate change, is the apotheosis of the idea that "everything relates to everything else." The issue arises fundamentally out of the growth of human population and wealth, out of industrialization, technological change, dependence on energy, an increasing standard of living, the striving for economic growth, and the many other aspects of a burgeoning global population that needs to be housed, clothed, and fed and that aspires to an improving quality of life. It is a result of normal, not aberrant, human behavior involving uncountable, independent decisions in daily life by individuals, by industry, and by governments all over the globe.

Societies are increasingly familiar with the need to regulate and influence such dispersed decision making that has social effects; the issue of control of greenhouse gases, or of their effects, is not different in kind from many other issues in the management of risks and the control of the externalities of technology. The substance of the interactions, the number of actors, and the immense scale involved mean, however, that central economic and political interests within and among states would be directly involved in any measures to limit emissions, as well as in the effects of warming that does take place. Oil, gas, nuclear, and other energy industries; agriculture, water, and coastal interests; population policies (including those opposing abortion); third-world development objectives; industrial growth and competitiveness goals; national and international ecological concerns; and specific political commitments, such as tax policy, are only the most obvious interests at stake.

The breadth of the interests involved also greatly complicates bureaucratic politics within governments and among international organizations. Essentially all government ministries and international organizations have an interest in at least some aspects of the issue, either in the possible effects of warming or in measures for prevention or adaptation. Overlapping jurisdictions, differing agendas and priorities, varying knowledge and influence, and competition for budgets and power all conspire to complicate trade-offs and make agreement on policy difficult to achieve. Moreover, the relevance of so many aspects of daily life to global warming means that policies adopted to deal with specific activities are likely to have unplanned consequences in other areas; and it means that events in subjects far afield, such as political change in Eastern Europe or China, can have a direct impact on the ability to reach consensus on policies to cope with global warming.

INTERACTION OF LARGE, COMPLEX SYSTEMS

A second major characteristic of the issue is its dependence on the interaction of two large, complex systems: the planet's ecosystem and the human socioeconomic system. Large systems have large inertia; change is slow to occur and difficult to bring about. Natural changes in the ecosystem occur over periods normally measured in centuries, making reasonably consistent and predictable climate an important factor in the development of civilization and making gradual adaptation to climate change possible at the human as well as ecological level. The evolution of the socioeconomic system, once measured in similarly glacial terms, is clearly more rapid than in the past, in terms of both growth and structural modification. Nevertheless, change in major structural elements of the socioeconomic system, such as the energy-production and -distribution systems or patterns of urban settlement, continues to be characterized by long time constants.

The inertia and large size of the systems have several consequences. One is that large efforts are required to effect any change. The continuing growth in populations and GNPs, which increases the demand for energy and thus the burning of fossil fuels, implies that even to reduce the *growth* of emissions would be a substantial challenge to national policies. To hold emissions fixed or actually to reduce them would be much more demanding and costly.[26] The design of policies to achieve reductions is one of the more problematic, uncertain, and contentious aspects of the entire issue. Costs are exceedingly difficult to determine, and many assumptions are necessary about rates of technological change, the effects of taxes and other possible policies, and the extent of cooperation among nations. The analyses that have been undertaken typically show costs that are small or possibly even negative (because of the cost benefits of improved efficiency and conservation) for reductions in greenhouse-gas emissions of 10 percent to perhaps 40 percent, to costs of hundreds of billions of dollars, or more, for major reductions.[27]

Another consequence of system inertia is that policy intervention requires a long time horizon, with changes in either of the fundamental systems coming about only slowly. The momentum in the systems as a whole guarantees that there already is a "commitment" to atmospheric change, whatever the magnitude of that change turns out to be and whatever policies are put in place. But it also means that action delayed will extend the change farther into the future and will increase its magnitude, since the effluents will continue and grow, while the systems are not likely to become any more amenable to manipulation. In addition, the most important gas, CO_2, has a residence time of about a century, guaranteeing effects well into the future from gases already released.[28]

Evaluations of policies intended to reduce emissions are as a result particularly difficult. The present value of benefits to be realized decades in the future is vanishingly small at any standard discount rate.[29] But even if the discount rate were assumed to be zero, allowing direct comparison of the future with the present, the costs and benefits of ill-defined future consequences are necessarily much more uncertain than the costs and benefits of specific current actions. Thus, it is predictable that the assessment of specific current policies to reduce greenhouse-gas emissions is likely to center on the short-term costs of those policies and on the opportunity costs of denying funding for alternative needs, rather than on the more hypothetical and controversial long-term consequences. To put it another way, the higher the uncertainty of future costs and benefits, the more likely the future will carry less weight in comparison to the present—a conclusion made more meaningful in this situation because of the large immediate costs that would have to be incurred, with no immediate benefits visible.

Ironically, the complexity of these large systems also suggests they could be vulnerable to dislocation in unexpected ways. Complex systems that are incompletely understood may have unsuspected nonlinear responses when subjected to unusual stress; for example, a major war can produce large discontinuities in the socioeconomic system. The ecosystem is resilient and self-correcting under a variety of major insults, but it is now being stressed by human activities that are increasing at a greater rate than ever before. Its possible responses to that stress are essentially unknown, making confident prediction of stability hazardous.[30] Surprise—such as the totally unexpected discovery of the Antarctic ozone hole—cannot be ruled out on the basis of present knowledge.[31]

UNCERTAINTY

A third characteristic is the role of uncertainty, an inherent aspect of all policy issues. In the case of climate change, the uncertainties are particularly large, contentious, and persistent. The policy process in principle needs information at several different levels. The first is information on the phenomenon itself—its causes and likely evolution. But that is of little policy interest without knowledge of the effects on, for example, precipitation, sea level, crop environment, ice cover, and frequency and violence of storms, as well as knowledge of the temporal and spatial distribution of these effects. At the next level, assessment of the larger consequences for agricultural production, prevalence of disease and pests, urban areas, coastal zones, population migration, economic development, energy demand, and international status, among other matters, becomes important. Along with this information, the policy process needs analysis of the alternatives for influencing the rate and extent of warming and analysis of the possibilities and costs of adapting to warming if it occurs. At each level, the uncertainties are likely to be larger, as the relevant variables are more numerous and future effects less well defined. The difficulties of dealing with long time horizons and the possibility of surprise already discussed add to the uncertainties that must be taken into account.

Removing or reducing all these uncertainties is not essential; policy is often made in the absence of definitive "downstream" analysis. However, if policies are particularly costly, reaching closure on particular courses of action will be much more difficult when affected interests are able to use uncertainty as an argument to prevent or defer action.

GLOBAL NATURE

A fourth major characteristic of the issue of global warming, one of particular relevance to this study, is the global nature of its causes, its effects, and the actions and policies required to modify it. No nation can solve the problem on its own, and the actions of one can be negated if others fail to act. The United States at the end of the 1980s was the largest producer of CO_2, at some 22 percent of the world total, down from 42 percent as recently as 1950. The former Soviet Union was the second largest at 18 percent; China was third at 10 percent, with a faster rate of increase than the others. Industrialized countries accounted for some two-thirds of the total, but the emissions of developing countries are growing more rapidly and will quickly become the dominant source.[32] That does not mean that unilateral action is irrelevant, only that it is not a solution.

The recognition that action by many nations is required may be useful to promote a global approach, but it can also be a deterrent to unilateral action if reaching international agreement proves to be difficult and contentious, or if a nation believes it can be a "free rider"—benefiting from the actions of others without bearing any of the costs. Moreover, the already-complicated bargains that have to be reached within nations to achieve policy consensus will be rendered even more complex by the requirement for international agreements, which are not likely to parallel those domestic bargains. Perhaps more significant in the long run is the absence of any institution in the international system with the authority to impose or enforce trade-offs as national governments are able to do.

The most politically troubling aspect of the global nature of the issue is the profound differences between the developing and industrialized worlds with regard to both causes and consequences. The emissions from developing countries will increase rapidly as a proportion of the world total as populations grow and development proceeds and as some, especially China, rely on their large coal reserves. Those nations are not likely to be willing to compromise their objective of economic growth in response to a problem that has been created by the previous behavior of rich countries. As developing nations are fond of pointing out, with justification, it is the industrialized countries that have benefited from using the atmosphere as a sink for the carbon by-products of their much wealthier societies, and that now propose limits on the use of that shared resource. Even though developing countries may have more to lose in the long run from global warming, since agriculture is a climate-sensitive industry and they are typically agriculture-intensive countries, the issue is not likely to be high on their immediate agenda. In any case, agreement on their part to limit emissions would require the infusion of money, technology, and skilled labor.

Though there are many subtleties to the issue, the essential conclusion is that to bring the developing countries along in an attempt to cap or reduce emissions will require large net additions to the transfer of resources from industrialized to developing countries.

It is also possible that some nations will see themselves as gaining if warming occurs, calculating that they will benefit from longer and more-productive growing seasons or from a more habitable climate. Or they could believe that the costs of warming will not be significant to them and in any case will be lower than the costs they would have to bear to help prevent global warming. Such nations might be expected to be unenthusiastic participants, at best, in negotiations toward international common action. Ironically, reducing uncertainty about the effects of warming might show that some states *would* benefit, thereby undercutting their interest in an international agreement.

The global character of the issue also has important implications for the need for functions to be performed in an international setting. The necessity of binding international agreements raises a host of difficult issues. On what basis are reductions to be allocated among nations? By GNP? By GNP per capita? By performance on energy efficiency? By local availability of alternative fuels? How will the difference in the effects of the various greenhouse gases be taken into account? Will differential dependence on foreign sources of fuel be factored in, and how will the extra foreign exchange needed for imported alternative fuels be provided? Who will incur the costs of conversion or other technological input, or of capital, or of reforestation? Who will provide technical assistance, and who will pay for it? Will information about technology be freely disseminated, and, if so, how will its development be financed? Will there be compensation for the reduced demand for some natural resources? Will the cuts, on whatever basis chosen, benefit some countries more than others?

If agreement is reached, how will it be monitored and enforced, and how will modifications be made as new information becomes available? What institution or group of nations will take on this task, by whose authority, and with what means to ensure compliance and effectiveness of operation? How will disputes be dealt with, and how will defection be handled?

International organizations have a substantial record in carrying out functions of allocation, monitoring, adjudication, and enforcement, and they perform more effectively than is often believed. Nevertheless, nations are not (or not yet) about to delegate significant powers to an external body over matters central to their political and economic interests. For now, it will be the basic nation-state structure of the world's political system in which policies will be considered and within which negotiations and implementation of whatever agreements are ultimately reached will be carried out. It is the contending policy processes of nations groping for agreement that will finally determine what is done.

DEPENDENCE ON SCIENCE AND SCIENTISTS

A fifth special characteristic is the near-total dependence on the theories, research, and calculations of scientists for assessment of the phenomenon—in fact, even for the awareness that there may be a problem. The theory of the possible effects of an increase of CO_2 in the atmosphere dates from the nineteenth century, but data on the accumulation of CO_2 and other greenhouse gases is much more recent.[33] Indeed, comprehensive data showing a sustained buildup of CO_2 in the atmosphere began to be collected only in the 1950s.[34] That data is growing and becoming more precise, and imaginative paleoclimatic research is showing that there was a relationship (not necessarily causal) between the greenhouse-gas content of the atmosphere and swings in the planet's average temperature over the past 160,000 years.[35] As of 1992, however, there was no decisive evidence that a warming trend due to rising concentrations of greenhouse gases was under way, though there was a rough consensus among many scientists of a possible half-degree temperature rise over the last one hundred years; the particularly warm decade of the 1980s seemed to justify the possibility.[36]

The word of scientists alone is sometimes enough for a hazard to be accepted—as with, for example, the unseen relation between a chemical and cancer. But there is usually accompanying hard evidence of some kind to justify corrective action. In this case, the hard evidence is lacking not only about the likely effects of a temperature rise, but even about the temperature rise itself.

Moreover, the forecasting of future temperature change, let alone its effects, cannot be done solely on the basis of gathering empirical evidence. Forecasts require understanding the outcome of the exceedingly complex interactions of the many variables that constitute atmospheric and ocean systems. The only approach to analyzing the functioning of the systems and their response to changing anthropogenic inputs is through elaborate computer-model simulations of the interactions. These large models, of which there are only a handful in the world, are designed by a relatively small number of scientists residing almost entirely in a few scientifically advanced nations.

These scientists would be the first to admit that their models are necessarily imperfect; that the data is incomplete, especially with regard to the oceans and the atmosphere/ocean interface; that potentially significant feedback effects (such as those due to clouds, or changes in absorption of CO_2 by the ocean, or nonlinearities in the system) are not now acceptably dealt with; and that the capabilities of computers are still inadequate for the scale of computations that must be run. They would also acknowledge that important inputs to their models—the rate of production of greenhouse gases in the future, the timing of CFC phaseout, the rates of economic growth—are social factors that can only be estimated or assumed. And most would admit that there is a considerable degree of incestuousness within the scientific community, of common paradigms being reinforced in intensive interactions among only a small group of researchers.

Nevertheless, the authoritative studies that have appeared reflect a consensus—largely but not universally shared—that has developed in that community: that the consistency in the predictions of various models, along with corollary evidence, provides a reasonable basis for concluding that the buildup of gases will have the temperature effects presented earlier, within the specified error bands. Few scientists would assert that the effects *will* definitely occur in a particular time frame, but many leave no doubt that they believe they will.

In essence, a majority of the members of a small community of scientists has reached a judgment, based on limited evidence and on imperfect models whose assumptions and calculations are not accessible to laymen, that has massive implications for the health of the ecosystem and for the fate of people and of nations. And the international political community has accepted the warning of the scientists in substance, if not in detail. The image of an inverted pyramid comes to mind, of a steadily broadening body of implications that rests ultimately on the point of a relatively small band of dedicated scientists who recognize the limitations and uncertainties of their work.

The dependence on scientists for the basic information is decisive in this issue in ways that have few parallels—policy issues associated with nuclear weapons systems may be the closest example. The uncertainties, however, make it inevitable that respected scientists will be found with sharply divergent views on the issue and its components, for there will be differing judgments about the importance of the many shortcomings in the data, the models, the interpretations, and the expectations of what future research will show. Whether or not there is a consensus among a majority of the scientists working in the field, there will inevitably be scientific experts testifying in opposition to one another and available to assist interested parties anxious to prevent or defer action—or for that matter, to accelerate action prematurely when the hard economic and political choices must be made.

THE PLANETARY THREAT

Finally, the "planetary" characteristic of the issue may be quite important. Climate change poses, or is seen to pose, a danger for the entire planet comparable, so far, only to the threat of all-out nuclear war and, in ecological matters, to the destruction of the ozone layer. Is that characteristic a significant factor in the emergence of ecological consciousness on this issue? It is an open, but possibly critical, question; for if costly policies are necessary, they will require broad public support to be enacted.

Consequences for International Politics

The scope and implications of this issue may be unique in international affairs; global warming may have significant political effects even as a potential but unproven danger, and possibly very far-reaching consequences if the forecasts prove to be correct. The effects will vary with the level of threat posed by the issue as more is learned; there are three relatively distinct levels.

GLOBAL WARMING AS AN EMERGING, UNCERTAIN QUESTION

As long as the extent and timing of warming and its effects remain uncertain and contentious, the implications for international affairs are not different in kind from those that arise in the context of interdependence, as discussed in chapter 4.

That is, global warming will become a more prominent focus of attention in governments and international institutions, and a subject for international debate and negotiation. The interdependence of interests will generate domestic debates about alternative policies, and political and economic interests within countries will join forces with like-minded interests in other nations to attempt to move governments in one direction or other. Additional resources will be committed to research to reduce uncertainty, better estimate costs, and develop technological options as well as economic and other policy options. The media, scientific organizations, and innumerable nongovernmental organizations will become involved in discussing the state of play or in promoting particular policy proposals. All of this was well under way in the early 1990s; the 1992 UN Conference on Environment and Development was seen as a milestone in the negotiation of a "framework" convention on climate change (one that sets general goals, leaving details for later negotiation).

The effect on international affairs, as long as global warming remains so uncertain, is essentially to add a new and arresting issue to international politics—one that raises particularly difficult questions for governments because of the many ramifications that have been described, but not one that is inherently unfamiliar.

This conclusion may seem to be contrary both to the extensive public attention that has been given to the subject and to the call by some political leaders and governments for commitment to substantial reductions in greenhouse gases.[37] However, unless overwhelmed by strong and enduring public consensus or by political leadership not yet in evidence, the political processes within and among nations attempting to come to grips with this issue are not likely to result in substantial action until uncertainties are greatly reduced, and probably not until there is palpable evidence that warming is occurring.[38] This is so even though there is a possibility of irreversible effects on the earth's climate if the forecasts of substantial warming are correct or underestimated; or if there are damaging surprises.

The policy processes of all nations will have difficulty dealing with this issue decisively; but among industrial countries, the policy processes of the U.S. government may prove to be the most resistant to formulation of comprehensive policy and early action. The divided governmental structure, with conflicting bureaucratic and legislative goals and overlapping jurisdictions and agendas, makes agreement difficult even on less demanding issues. Virtually every federal agency has a legitimate interest in the issue; correspondingly, most congressional committees are engaged, each with its turf to defend or expand and with a necessarily limited vision of the problem as a whole. Conflict between the overlapping jurisdictions of House and Senate committees, and between Congress and the fragmented executive branch, further complicates the matter. Coherent policy outcomes are difficult to achieve, especially policy outcomes that would, despite uncertainty, mandate significant costs for powerful segments of the society or that would conflict with entrenched political commitments in matters such as taxes, abortion, or even the appropriate role of government.

The inherent uncertainties in the issue, it should be emphasized, extend not only to the dimensions of the warming and its effects, but to the costs of mitigation or of adaptation to the warming if it materializes. The costs of future adaptation for some countries, especially the richer ones, may well not be very large, even in direct comparison (without any discounting for the future) to the costs of prevention. The National Academy of Sciences study concluded, "People in the United States likely will have no more difficulty adapting to such future changes than to the most severe conditions in the past, such as the Dust Bowl."[39] Given that view on the part of influential analysts, it becomes quite unlikely that actions with substantial short-term costs will be undertaken even in the wealthier countries, at least until much more is known.

Some governments, particularly those in Western Europe, have publicly committed themselves to action on the issue, typically setting targets to "stabilize" greenhouse-gas emissions at 1990 levels by the turn of the century or shortly after.[40] These are symbolically important commitments, though not particularly costly; nor were the measures to achieve them fully fleshed out at the time they were made. The United States has resisted such a formal commitment. But political pressure for the United States to agree to some reduction in emissions has become severe in preparation for the 1992 Conference on Environment and Development in Brazil.[41]

It is possible, of course, that the strength of evolving and widespread public attitudes toward protection of the environment, the dramatic planetary nature of the issue, and the accompanying growth in organization and influence of national and international environmental movements could lead governments to make much more substantial commitments to reduce greenhouse-gas emissions. The Greens have become potent political forces in many nations; in Eastern Europe, as we saw in chapter 4, concern over the almost-total neglect of the local environment was one of the factors that contributed to political change. Undoubtedly, public sensitivity to the environment will be of growing importance; but it remains problematic, at best, whether environmental movements can decisively influence publics and national policy processes when costs are large, other needs pressing, uncertainties high, and interests affected so diverse.

The successful conclusion of the Montreal Protocol to eliminate production and use of CFCs because of their destructive effect on the earth's ozone layer might indicate otherwise. That agreement is, in fact, an important precedent, an example of international action taken before certain evidence of damage was in hand; but it also serves to underscore difficulties. The ability to negotiate and ratify the agreement was aided by the dramatic evidence of the unpredicted ozone "hole" over the Antarctic (a form of direct evidence), while the immediate economic costs of regulation were relatively minor. Negotiating the initial agreement, before the cause of the Antarctic hole was manifest, proved to be quite hard and resulted only in a commitment for a 50 percent cutback of CFCs. Subsequent negotiation, when the evidence of the effects of CFCs was clearer, led to a decision for an essentially complete phaseout of production by the end of the century.[42] The experience of the negotiations on CFCs will be useful for dealing with global warming, but the two issues hardly compare in complexity and breadth; in fact, the extent of difficulties encountered in banning CFCs when so few interests, relatively, were at stake does not offer much encouragement.

The importance of "hard" evidence as an energizer of public opinion is particularly relevant to the question of the staying power of public attitudes. The heat and drought of 1988 in North America, coupled with a scientist's testimony before Congress that tied them to the greenhouse effect, helped to catapult the issue high on the environmental agenda, at least in the United States.[43] Though scientists believe that a causal connection cannot be scientifically demonstrated, the testimony nevertheless served to catalyze attention and opinion. Undoubtedly, several more summers of that kind, whether there is a relationship to average global temperature or not, will continue to fuel public pressure and make it possible to surmount obstacles in the policy process. On the other hand, a series of cooler-than-normal years quite unrelated to overall climate change might decrease public interest and cut back the potential for political mobilization. Even if there continues to be general public concern for the environment, the intensity of the concern about climate change is likely to vary with the transient vagaries of weather and climate.

GLOBAL WARMING AS AN ACCEPTED THREAT

The situation could change materially if there was compelling evidence of a temperature rise that led to acceptance of the reality of the threat of global warming, and, importantly, of the serious danger and costs of its effects. It would then be easier to mobilize a political consensus for measures to reduce or counter greenhouse-gas emissions.

Those measures would have to be substantial in scale and cover a wide range of subjects. They would include international agreements mandating reductions in emissions, with complex arrangements for national quotas, allowances for a varying mix of gases, credits for energy efficiency, transfer and ownership of technology, financial support for alternative fuels or technologies, possibly an elaborate tradeable-permit plan, and many other matters. International means would be required to monitor compliance, maintain an up-to-date data base, manage resource transfers, operate international financing and research, mediate or adjudicate disputes, allocate resources, impose sanctions for violations, and deal with the particularly difficult problem of free riders. International agreement might have to be sought to consider proposals for engineering solutions that would reduce incoming solar radiation on a global basis (for example, by artificially injecting particulate matter in the atmosphere or by altering cloud abundance through cloud seeding), proposals that would raise hopes for an inexpensive technological fix or fears of dangerous environmental meddling and unanticipated side effects.[44]

In these circumstances, several elements of international politics would take on new dimensions. In particular, the level of international cooperation among states would have to be considerably more intensive than it has ever been, with intrusive agreements necessary that would affect central aspects of national economies, especially industry, energy, and transportation. Substantial transfers of resources from North to South to ensure the participation of the developing countries would be required. And more authority would have to be delegated to international mechanisms than states have traditionally been willing to allow. The autonomy of national governments with regard to the setting of policy within their borders would be significantly affected by that delegation of authority and by the need to conform to stringent agreed international norms.

Perhaps more important, the political influence of various interests within a nation in *both* foreign and domestic policy would be curtailed by the imperative of reaching binding international agreements. In effect, there would have to be greater centralization of power within the state to prevent competing interests from frustrating either the negotiation of agreements or their ratification and implementation.

Politically stable states with effective policy processes are likely to be able to develop adequate responses in this situation. States with less secure political foundations, or those that may be politically stable but have policy processes vulnerable to gridlock, will find agreement harder, perhaps impossible, unless genuine catastrophe looms or an external power intervenes. The need for a hegemonic power, to provide the combination of leadership and coercion essential to bring about action in recalcitrant states, may become overwhelming.

The international machinery required for implementing agreements need not consist only of existing or new intergovernmental organizations. It could also include processes set up and managed by individual countries or even by nongovernmental organizations, by agreement of all. A hegemonic power, for this purpose as well, may prove particularly useful to orchestrate international action and to take on politically sensitive management responsibility that requires the authority bred of power.

The increase in authoritative responsibilities that would, through whatever form, have to be exercised in an international setting would be substantial. To carry out those necessary, high-value political functions, a meaningful transfer of power and authority from individual states would be required—either to international organizations or to other states.

Of course, whether these speculations become reality or not depends on the perception by governments and publics of the severity of the effects that would result from warming and on the estimates of the costs of prevention and adaptation. The difficulties and costs of implementing measures to abate emissions of greenhouse gases, even if the predicted threat is moderately severe, may lead states, particularly the richer nations, to prefer a policy of adaptation rather than prevention. That is the more likely outcome unless the threat is seen as being clearly of great cost to all, or unless the effects of warming show up in the form of a dramatic local or regional climate change that precipitates widespread public outrage and that can be demonstrably attributed to greenhouse-gas buildup.

Under this assumption of the reality of the threat of global warming, there would be some unavoidable increase in average temperature as a result of the greenhouse gases already released or the inadequacy of preventive actions. That warming would have effects with direct consequences for international affairs. Those consequences are not predictable in detail, since the amount of warming and its effects are unknown, but the implications that would likely follow are reasonably clear: dependency relations would be altered as a result of changes in patterns of production of food and essential commodities; the economic status of states would be affected as sources of wealth and costs of maintaining living standards changed; the scale of migration of dispossessed or destitute populations would grow as living conditions in some countries deteriorated; and, as a result of these and other effects, the economic balance among nations would be altered as some nations were found to benefit (or to lose proportionally less than others) as the effects of warming unfolded.

Such changes would most likely be less dramatic among the wealthier nations, which would be better able to accommodate to the new circumstances, and correspondingly more significant among the poorer countries. It is possible that the beneficial effects for some countries in northern latitudes (better growing conditions, expansion of arable land, easier access to natural resources, improved transportation and living conditions), together with the negative effects for others, could be so substantial as eventually to alter existing geopolitical relationships. Even with dramatic alterations in the status of some nations, however, the international political changes to be expected even from appreciable warming (barring major ecological surprises) would likely be in the roles of the players and the substance of the issues, not in the underlying structure of the system.

GLOBAL WARMING AS A CATASTROPHE IN THE MAKING

The situation would be different again if there were to be a strong consensus in the scientific community of genuine global catastrophe in prospect if warming were allowed to continue, and if signs of such a catastrophe began to appear: for example, visible evidence that the Gulf Stream might drastically change course; or the first signs of a breakup of the West Antarctic ice sheet, which would threaten to raise global water levels by five to ten meters; or indications that a runaway greenhouse effect, akin to what presumably happened on the planet Venus, was genuinely possible.[45] All of these have been considered as possible, if unlikely, consequences of global warming. They take on more than casual plausibility because of the fact that surprise and nonlinearities in the response of the

ecosystem to warming can be neither predicted nor ruled out. The stressing of the system at the present unprecedented rate will almost certainly
produce some surprises, though not necessarily on a catastrophic scale.

Such a pending global crisis would put the international political system
under great stress, with results that are as hard to predict as the ecological
surprise itself. There are several possibilities of how the overwhelming
common threat might be met: an entirely new level of cooperation among
states and sacrifice of autonomy could emerge; a single state or small
group of states could decide it was necessary, through bald use of power,
to dictate concerted action; the UN or another international organization
could be granted sufficient authority and resources to force action and
compliance; or some combination of these could evolve.

Each of these possibilities, with its many shadings, is a variation of one
of two dominant schematic paths the international system could take: the
submerging of nation-state identity in a common global framework, or
submission to the will (and power) of one or a few dominant states. In the
first case, the political evolution from the current international system
would certainly be aptly described as fundamental; there would be a new
world system of shared power and limited state autonomy. The latter path
would be simply an assertion of the traditional politics of power and the
rule of force, involving the emergence of a state or states with clear hegemonic dominance.

A world crisis on this scale may be the only imaginable route to substantial, rapid evolution in the international political system, but it is a risky
route. The response to the crisis could fail; or it could, even if successful
in warding off calamity, lead to political outcomes that proved to be considerably less desirable than the imperfections of the existing system.

These three levels of the development of the threat are not, of course,
mutually exclusive or distinct. They overlap considerably, and, correspondingly, the political responses to them will not be as distinct as implied here.[46] Unless the threat of global warming moves at least to the
second level, however—and quite possibly not even then—the consequences for the political structure of the international system are not
likely to be substantial.

Additional Thoughts

Broader philosophical questions raised by the threat of global warming
have been intentionally avoided in this discussion so far. Yet many believe
that the planetary nature of the threat can carry great moral as well as
practical significance and that it ought to be the overwhelming consideration. They argue that humankind has no right to ignore matters that could

put at risk the only planet in the solar system on which life has developed, or at the very least that planetary-scale effects deserve a different calculus of costs and benefits than more prosaic concerns.[47] Some see the question of intergenerational equity as a central one; they maintain that the environmental welfare of future generations deserves a direct entry on the balance sheet that assesses prospective actions.[48] Others, more generally, see the preservation of environmental quality as a moral imperative, essential if the human race is to retain its humanity.[49]

These are all powerful beliefs, but are difficult to reflect adequately in a policy process. In the formulation of policy, qualitative factors are typically dominated by quantitative assessments of costs and benefits and by carefully targeted pressure from affected interests. The planetary character of the threat may, however, be of sufficient emotive power that it will carry considerable psychological weight, adding substantially to the policy impact of new information that makes the threat more believable and the prospect of damage more disturbing.

The fundamental significance of population growth to this issue must also be noted. We discussed population growth as an aspect of the evolution of international affairs in chapter 4, but it has here a particularly critical role in increasing the sources of stress on the global ecosystem.[50] Much of the growth is taking place in the developing countries—those least able to allocate resources to reduce environmental impact, and also those least concerned with long-run global issues in the face of their immediate problems of hunger and poverty. Yet continued growth of population necessarily will result in increased emissions of greenhouse gases and other environmental consequences, as individuals seek food, heat, and shelter and strive to improve their material well-being. Eventually, the growth in population must be brought under control, for global ecological reasons as well as for many others; the longer the delay, the larger the impact and the overall costs to the system. It is a central issue in many international concerns; it is a central, though often sidelined, issue in this one as well.

Lastly, it should be pointed out that the threat of global warming is contributing to recognition of a broader definition of national security. A nation's security, long considered the most basic responsibility of government, is gradually being accepted as having many more dimensions than its traditional military focus. Economic strength, educational quality, health of the population, and scientific and technological capability are all now increasingly recognized as important components of security in their own right, as the threats to a nation are clearly more subtle and challenging than the threat of military force alone.

Richard Ullman, aiming to codify these broader dimensions, defines a threat to security as "an action or series of events that (1) threatens drastically and over a relatively brief span of time to degrade the quality of life for the inhabitants of a state, or (2) threatens significantly to narrow the range of policy choices available to the government of a state or to private, non-governmental entities . . . within the state."[51] Clearly, the nature of the threat posed by global warming could match that definition closely.

So far, however, military matters still dominate national-security policy agendas, and only the state of a nation's economy has come to be seen as approaching equal importance as a security issue. Given the many potential conflicts between economic growth and mitigation of global warming, the increased prominence of economic matters means that policies to prevent global warming are not likely to be accorded much standing in a national-security framework. That is unlikely to change unless the threat becomes an evident danger to the security and well-being of states.

Other Global Dangers

The greenhouse-effect issue has been as explored in some detail in this chapter as a representative example of a new class of issues emerging in international affairs. The prime characteristic of these issues is that they pose potentially large threats to the entire planet; their causes may be either natural or man-made, or both. Most issues that have such planetary significance involve massive environmental change, but there are threats arising in other subjects that qualify as well.

Several of these other global dangers are worth mention. One is pandemics—epidemics of serious disease occurring over a substantial portion of the planet. Acquired immune deficiency syndrome (AIDS) is an obvious example, a disease that has apparently spread worldwide largely through the availability of inexpensive intercontinental transportation.[52] The eventual extent and devastation of the disease are unknown, though it is clear it will have catastrophic effects in some developing countries in Africa, which will lose substantial proportions of their young, able-bodied populations. The Black Death is an earlier example that had devastating results in Europe in the fourteenth and fifteenth centuries.[53]

Pandemics affecting plants are also a possibility. The increased and widespread reliance on genetically similar seed strains (discussed in chapter 4) makes modern agriculture vulnerable to sudden, massive damage from disease or pests. In the past, such attacks were local or at most regional; now they could occur worldwide. Significant problems may arise in the oceans; some scientists, for example, are concerned about a "spreading global epidemic" of both toxic and nontoxic algae blooms, which have major economic impacts by choking marine life and fouling beaches.[54]

And there is a host of issues that are less immediately dramatic but that are genuinely global in their effects—effects that may not be reversible and that may, over time, lead to serious consequences. Destruction and loss of biological species, depletion of the living resources of the sea, deforestation, and desertification are some of the present concerns.[55]

A last category is threats from outside the planet. The earth had a close call on March 23, 1989, when an asteroid traveling at 46,000 miles per hour passed within 500,000 miles of the planet, crossing the earth's orbit a scant six hours after the earth had passed. That is a remarkably close encounter as reckoned in interplanetary distances. It was estimated that, had the asteroid collided with the earth, its effects would have been equivalent to the explosion of a thousand one-megaton hydrogen bombs.[56] That interplanetary object was detected only after it had passed. Another, smaller asteroid, which was detected twelve hours in advance, came even closer—170,000 kilometers (100,000 miles), closer than the moon—in January of 1991. Had it hit the earth, the impact would have had the explosive power of an estimated forty kilotons of TNT.[57] It is hypothesized that it was an impact with an extraterrestrial body that led to the demise of the dinosaurs.[58]

Eventually, it will be possible to give warnings of impending impacts of such extraterrestrial objects, raising the possibility of cooperation in the construction of a massive system for detection and, presumably, deflection through the use of powerful, almost certainly nuclear, explosions.[59] Though the probability of encounters with such objects may be very small and the costs of preparation to intercept them high, the consequences of a collision would be serious enough that there will likely be substantial pressure for action once meaningful action becomes feasible. Cooperation on such a system would, among other implications, raise questions about the design of missiles and explosives that would be equally useful for warfare; such a project would be a major test for the international system and would undoubtedly lead to significant long-term political effects.

These threats involve issues and consequences conceptually similar to those encountered in the example of global warming, though obviously the details differ. All but the possibility of asteroid impacts have become global threats as a result of human activities, whether the dangers arise from the side effects of meeting the demands of population growth and economic growth, or from the spread of disease vectors through easier and more-widespread movement of people and plants.

The materialization of these dangers, or attempts to prevent them or limit their impact, would likely have effects on the international political system similar to those in the case of global warming. There would be substantial new constraints on the autonomy of nations as international agreements to curtail specific national activities were of necessity forged and implemented, with resulting tendencies toward the centralization of power. Cooperation among states would be materially expanded, and either the authority and activities of international machinery or the power of dominant states would grow to be able to take on more extensive and more politically sensitive tasks. Resources and technology would probably have to flow from the wealthier to the poorer countries to ensure adequate participation by all. Ultimately, the nature of an issue and the severity of the danger it posed would determine how difficult these developments would be and the nature of the change that would take place in the operation of the international system.

Some of these global dangers are closely related to one another; some are largely independent. But it is worth noting that the experience of the international community in meeting any of them will have an impact on how well others are dealt with. For that reason, the first major item clearly on the agenda—the destruction of the ozone layer—must be tackled quickly and with vision. Wisdom shown in overcoming that danger will set important precedents for dealing with global warming and with other threats as they arise.

New dangers not yet thought of are sure to emerge, and issues on the agenda today will develop in unexpected ways that involve discontinuous and synergistic relationships. Surprises are inevitable. And they will be increasingly likely as stress on the ecological system continues and intensifies as a result of human activities that go far beyond previous experience.

How the international political system can cope with the set of issues these new dangers represent—and whether it can—is far from clear. For threats that remain sufficiently uncertain, the response is likely to be along familiar lines. If the dangers become real and immediate, substantial evolution in international affairs will be required that could either reinforce old patterns of hegemonic dominance or establish new systems of extensive cooperation and sacrifice of national autonomy. Which pattern would emerge remains an open question.

Six

Practical Problems of Governance: Institutions and Processes

A FREQUENT theme in the preceding chapters has been the effects of technological change on the nature and problems of governance within countries and internationally. There have been many references to the new demands placed on national policy processes, the altered setting in which foreign policy must be made, and the additional international functions that must be carried out through collective action. These consequences for governance are in themselves factors influencing the evolution of the international political system; it is important to pull the various strands together and examine them as a group—what they are and what their influence may be.[1]

Time

One of the most interesting developments that accompanies technological change is the greatly enhanced significance of time. That is evident at both ends of the spectrum: the time available for consideration of complex decisions, sometimes life-and-death decisions, can be vanishingly small; while, concurrently, governments have to make policy for material issues with time horizons extending to many decades and, sometimes, thousands of years. Both phenomena are essentially new as significant factors in the councils of government; both have proven to be exceedingly difficult to accommodate in policy processes.

The compression of time scales was evident in the discussion of several sectors; the most dramatic case was the implications that modern strategic-weapons systems have for command and control of forces (examined in chapter 3). The brief delivery time for highly destructive strategic missiles (thirty minutes or less), combined with the massive quantities of data from real-time reconnaissance and warning systems that must be processed to determine whether an attack is in fact under way, leaves little time for considered review of policy options. Responses must be largely prepared in advance, all but foreclosing the exercise of human judgment in the context of the actual situation.

The proposed U.S. Strategic Defense Initiative, were it ever operational, would pose the problem even more starkly, for there could not even be the fig leaf of required referral to the president for decision. Any attempt to destroy missiles during their launch phase would require the firing of antimissile weapons within three minutes, when the offensive missiles would still be in the atmosphere (though even that time could be further reduced). That would necessarily mean a system with automated, procedural responses to electronic signals, allowing no concurrent human evaluation or intervention.

The problem is evident in conventional-weapons technologies as well. The increased speed of many weapons systems, the much more complex information environment, and the damage that can be caused by individual weapons can severely limit the time available for considered decisions about appropriate response.[2]

Time compression is also evident in economic sectors. For example, the growth of global, twenty-four-hour computerized financial markets (discussed in chapter 4) has resulted in a trading system involving literally trillions of dollars that is dependent for its operation on computerized evaluation of market developments and on preprogrammed trading decisions. In this situation, governments must be able to respond with great alacrity (not a common characteristic of governments) when incidents arise—such as excessive price volatility that leads to panic sell-offs—that would have serious economic consequences. Such incidents might themselves be a product of the instabilities introduced by computerized program trading and by the possibility of machine (computer or communication) failure.

The rapid diffusion of information worldwide through communication networks and the media leads to a different form of time compression: pressure for quick government reactions to important developments, wherever they may occur. Officials might prefer time for contemplation or analysis, but high-profile issues that emerge instantaneously on the global stage often require an immediate response. The absence of response or a deferral can have high costs, including loss of control of an issue. In any case, the time available for consideration of important policy issues is often severely truncated as interested parties at home and abroad are quickly informed and add their voices and actions to the pressure for hasty decisions.

The need for rapid response increases the reliance on the first information received about an event, information that is likely to be incomplete and woefully inaccurate. This enhances the risk of emotional or irrational responses on the part of either decision makers or the public, or both.

The situation is exacerbated by the broadening of the agenda with which senior government officials are involved. The time that can be devoted to any single issue becomes steadily more constrained, so that the availability of those officials becomes a critically scarce resource in the policy process (a development typically little appreciated outside government; senior officials are expected to be knowledgeable about, and have time for, any issue that arises).

The late Harold Macmillan, a former British prime minister, decried with his usual style what he called this "instant politics"—the assumption that leading officials should, because of the existence of radio and television, be expected to comment immediately on any event: "Every day they talk on television—every day, immediately an event comes. There is no time for thought, no concentration; you have to come on at once or else you're thought to be out of it. . . . You must get away, you must read a book. . . . You must be quiet."[3]

One consequence of this time pressure in the policy process, pressure that is greatest for those issues that pose the greatest dangers and are most likely to threaten a crisis situation, is the tendency to centralize foreign- and security-policy responsibility within governments. There may no longer be time for extensive consultation on an issue, and a head of government may not be willing to entrust responsibility to a subordinate official when considerations of time may make that equivalent to delegation of authority. Moreover, it usually will be, or will seem to be, more efficient, faster, and safer to conduct policy from a central office when rapid response may be required. Hence the tendency for security policy to be centered in and conducted from the office of a nation's leader.[4]

Another consequence is to increase the reliance on machines for the collection of data, for its analysis, and for the transmission of a response. Technology is not only the cause of the problems of compression of time; it is also the primary means for dealing with them. Technology makes it possible to reduce, process, and analyze information and to implement decisions much more rapidly than in the past. There is thus greater dependence on machine-generated data and analysis, as well as greater vulnerability to the unavoidable imperfections of the equipment and to the built-in biases and assumptions of those who design or program it.

At the other end of the spectrum, governments are grappling with major issues for which decisions must be made today but that require consideration both of consequences and of the viability of commitments extending far into the future. The danger of global warming (discussed in chapter 5) is a vivid example; measures to prevent or delay its arrival would have immediate effects on national economies, while the consequences of warming that did take place would appear not in the next year or two, nor in the next political cycle, but over generations. The disposal of radioactive or highly toxic waste is another issue that requires policies today that take account of the commitment to safeguards over thousands of years.

The ability to weigh policies appropriately in these situations is hampered by many difficulties, not the least of them the problem of comparing investments intended to produce benefits over varying times in the future. Cost-benefit analysis using any reasonable discount rates to account for the time value of money makes the present value of expenditures whose benefits will appear after thirty, forty, or more years usually uncompetitive with expenditures that have shorter-term payoffs.[5] The problem is exacerbated by the inherent uncertainty in the estimation of future benefits, because the immediate expenditures required to reap those uncertain benefits must be compared against expenditures for alternatives whose benefits are nearer and clearer. The result is that it is difficult to give substantial weight in the policy process to the needs of the future in comparison to those of the present.

Alternative approaches have been considered—for example, the use of a "social" discount rate in cost-benefit analyses that could be zero or even negative.[6] Given the intense competition for resources to meet immediate social needs, however, governments would not likely be willing or politically able to introduce procedures that would generally bias decisions toward future needs and against current ones.

Thus, the significance of time in the policy process in governments has taken on new and particularly difficult dimensions. This change must, however, be kept in perspective. Time has always been a scarce resource in the making of policy, always a constraint on deliberative formulation of policy. Nor is it all that unusual for governments to make decisions and substantial investments (dams or water-distribution systems, for example) whose benefits are realized well beyond contemporary generations or beyond the terms of the officeholders who made the policy.

The difference now is that, at the compression end of the spectrum, not only is time relatively an even scarcer resource than before, but more decisions, and more consequential ones, are physically required to be made in a very much shorter period of time. At the other end of the spectrum, long-term effects must be explicitly taken into account in a larger number of cases, with many of those cases having global consequences.

The disparate time scales of technological and societal change also add to the pressures on decision processes. Technology evolves faster than the social context and regulatory environment in which it is embedded. Even with efforts at technology assessment that attempt to anticipate the broad, unplanned effects of technology, the social setting will always be largely reactive to technological change as it appears. The result is that policy processes are continually faced with new, unanticipated situations stimulated by technological change and with new requirements for accommodating to its effects. The capability for rapid response to change becomes a significant element in a policy process—a process that is inherently structured for a much slower rate of change.

Technical Content of Issues

The inextricable and pervasive involvement of scientific and technological factors in so many issues in international affairs, a relationship we have seen throughout this study, means simply that those factors must be sensibly included in policy processes. How relevant they are to a specific issue—whether their details are central to its consideration or simply part of the general framework—will vary from issue to issue and within issues. In this respect, scientific and technological factors are no different from other aspects of policy: they are an element that must be considered but whose importance depends on the issue and on the particular questions being asked.

Policy processes, then, must include effective representation of the technical aspects of issues. Most governments, however, have found that to be much harder to accomplish in their foreign-policy establishments and at senior levels of government than might be assumed.[7]

One of the many difficulties encountered in all policy areas but particularly, it seems, in the foreign-policy domain is that many of those in foreign-policy positions have a limited understanding of science and especially of technology. Whether the problem is due to the recruitment and promotion patterns for foreign-service personnel, the traditions and nature of their roles, the necessarily limited technological resources at their command, or other factors, the result is often limited ability to consider anything other than the surface technical characteristics of issues as presented by the mission agencies of government. In turn, those agencies typically have only a limited understanding of the broader foreign-policy aspects of an issue, and moreover have a stake in presenting information so as to support their own policy objectives.[8]

In the foreign-policy process, the result often is inadequate capability to relate policy choices to national interests, implicit deferral by foreign offices to the policy perspectives of technical departments, misunderstanding by foreign-policy officers of the dynamic nature of technology and of its potential responsiveness to policy intervention, and, ironically, often a highly exaggerated view in the foreign-policy establishment of what technology might be able to achieve.[9] There have been many studies in the United States of this problem in its foreign-policy process, with numerous suggestions for change. These studies have had little effect on operational or staffing patterns in the Department of State.[10]

The overvaluation of technology's potential contributions is a relatively new phenomenon; it has perhaps become as much of a problem as not understanding technology's role in policy. The cause is obvious. Not only is advancing technology impressive in contributing to the expansion of wealth and capabilities, but science and technology have been increasingly used to multiply options for dealing with recalcitrant issues. Whether it be substitutes for scarce resources, more-productive crop strains to stave off famine, means for observing military movements to avoid surprise, or moon landings to refurbish international prestige, the necessary technologies have been developed apparently on demand.

But, as discussed in chapter 3, not all desirable technologies are feasible, nor are they all viable in society even if technically feasible. More important, whether or not technology is useful in resolving a significant social issue, it is certainly never the sole determinant of the outcome of the issue. Science and technology offer greater flexibility and a broader range of options for policy than were available in the past, but excessive optimism about the capability of technology to solve social issues must be severely restrained. Technology can contribute, but social problems require social solutions.

It is not necessary that those in foreign-policy positions become scientists or engineers in order to be able to work with the scientific and technological aspects of issues. Rather, a well-crafted program that offers the same level of exposure to science and technology that is encountered in good programs of economics, industrial management, or public health can provide the basic tools that are needed—including an appreciation of how to approach the scientific and technological factors of an issue, how to understand their interaction with other elements, and how and where to obtain further information and analysis when required.[11]

One increasingly common circumstance in policy processes does, however, pose large problems for laymen and even for those with technical training. It is the need to represent the technical aspects of an issue in a complex, usually computer-based, often opaque (to the layman) analysis. The internal structure of these analyses and the assumptions that must go into them are critical to what conclusions are reached, but they are not inherently independent of the personal views of the analyst: they require the exercise of personal judgment in the selection of the important variables, in the simplification that is always required, and in the choice of analytical methodologies. Policy conclusions or options are unavoidably vulnerable to the shading that results from the intended or unintended biases of those doing the work, but the biases are difficult to discern because of the technical complexity of the analysis. Whatever the significance of the bias to the conclusions, policymakers are dependent on inputs that they cannot personally understand or evaluate. This dependence is made more problematic by a tendency for undue credibility to be accorded to computer outputs and to quantitative, rather than qualitative, analyses.[12]

This dependence on complex technical analysis is not a wholly new problem of governance; it has been particularly evident in security and energy matters for some time and is increasingly apparent in regulatory and environmental subjects. But the significance of the technological aspects of a growing number of politically salient issues makes this a more common problem—especially at the higher echelons of government, as more issues with major scientific and technological components reach those levels.

There is no "solution" for this problem of dependence on technical analysis. The single most useful prescription is for senior officials to have advisers and staffs close to them that understand their policy objectives and that are capable of evaluating such analyses in depth, are able to integrate the technical and policy dimensions, and are in a position to seek alternative analyses when necessary. To be effective in this role, an adviser must have the important and rare attribute of being well-versed in both the policy and the technical aspects of an issue.

The subject of global warming, explored in chapter 5, provides one of the more striking examples of dependence on technical analyses, and is unusual because of the centrality of the role of science and scientific research. The forecast that global warming will result from increasing concentration of greenhouse gases is based on elaborate models of global atmospheric circulation that are analyzed by means of repeated computer runs. The models are complex, highly technical, limited in number (there are essentially only five major models associated with large computers throughout the world), and include assumed relationships among important elements that are not yet well understood (e.g., ocean absorption and the role of clouds). Yet the predictions are the justification for placing the issue on the global agenda, where it could have important political effects and potentially precipitate significant change in the structure of the international political system.

The role of the scientific community in this subject is remarkable. The computer modeling and analyses that are the basis for the concerns over warming are the product of the work of a small community of atmospheric scientists, working to the best of their abilities and judgments with the latest, but still-insufficient, tools. They must simplify the enormous complexities of the atmosphere in the design of their models, which even then are not accessible to the layman; and they must present results that can be no more than uncertain predictions, the importance of which is in turn a matter of judgment.

The responsibility shouldered by those scientists in their work—and, at least as important, in how they present it—is enormous. By extension, the organizations of the international scientific community that are now heavily engaged with the subject and its implications also bear major responsibility.[13] There are massive policy implications dependent on what these scientists produce and when. Perhaps the reliance on the scientists of the Manhattan Project represented a similar responsibility for that time; no other example comes close.

The extent of the dependence on science and scientists in the case of global warming may be unique, but the increasing relevance of science and technology to international policy issues mandates an important role in policy processes for individuals with scientific and engineering training. They may not be the "new skilled policy elite" that will dominate policy, as was once predicted, but they have a responsibility to be as much a part of the process of integrating technology with other relevant factors as do the individuals without technical backgrounds who must learn how to cope with the technological factors.[14]

Role of Foreign Offices

The intensified integration of economies and societies, one of the more dramatic international effects of technological change, necessarily means large increases in the sheer volume and breadth of relationships between and among states. We have seen the important general implications of this development for international affairs; it also has more-prosaic effects on policy processes, effects that tend to undermine the ability of foreign offices to maintain their traditional dominance in the making of foreign policy. The portions of government formerly concerned almost exclusively with domestic affairs—agriculture, industry, health, environment, and education, for example—are increasingly engaged in matters that directly affect international relations and a nation's foreign policy. Those "domestic" departments have become legitimate players in the making of foreign policy; foreign offices no longer can automatically dominate the process, as it was once assumed they should.

The result is that, for a wide variety of foreign-policy issues, the foreign office is now typically but one among many ministries involved; the relative influence of each is determined by the particular circumstances and personalities concerned.[15] The increasing role of "domestic" ministries in international affairs cannot be halted or reversed; it is a characteristic of the intensely interdependent world that is continuing to develop.

Attempts to reassert the authority of a foreign office, typically by moving decision-making processes higher up the organization chart in order to engage the foreign secretary directly on a larger range of issues that have foreign-policy implications, cannot be continued indefinitely without overburdening the time of senior officials. Ultimately, such moves simply re-create lower-level problems at a higher place in the government hierarchy.

This general problem of formulating policy and coordinating governmentwide activities that affect international relationships is exacerbated in the United States, where power is divided between the legislature and the executive. Policy processes are more chaotic in the Congress, so that coherent national policy positions are even more difficult to develop than in typical parliamentary systems.

The dispersion among agencies of involvement in international issues would be enough on its own to bring about the gradual loss of dominance by foreign offices. The role of domestic technical agencies, however, with their greater knowledge of relevant technologies, is enhanced by the increasing importance of the technical aspects of issues. Though we have been at pains to point out that technology is never the sole determinant of the parameters of an issue, in the policy process knowledge is power, especially if there is a significant imbalance among participants in their understanding of a key element of an issue. When technological aspects are critical, the domestic agencies' greater technological knowledge makes it possible for them to present those aspects in ways that support their view of the international interests of the nation (and of their own interests) over the view of the foreign office. It is difficult for a foreign office to counter such slanted presentations, or sometimes even to know there may be alternative interpretations.

International Cooperation

The requirement for collective action to achieve common purposes or attack common problems is a natural consequence of the increasing integration of economies and societies. That requirement is bound to grow as integration increases and becomes an ever more prominent aspect of international transactions among nations. Much of the existing international cooperation takes place in the private sector, without extensive government involvement; a large and growing portion of it requires the commitment and agreement of governments on subjects of substantial political and economic significance.

The process that determines the policy of a nation toward collective action, however, continues to be entirely national in structure, giving representation to domestic interests affected by the issue and only indirectly to foreign or international interests. At the least, the process skews the consideration of international issues by underrepresenting the stakes of affected foreign parties or of the broader global community; at worst, it results in a parochial, nationalistic view that gives little weight to broader concerns.

In these nationally based processes, politicians or officials find it easier and safer—rational, in fact, from their perspective—to choose to take short-term gains through unilateral domestic actions rather than to gamble their own interests on the usually messier problems of international coordination or cooperation.[16] Moreover, in nations with open political structures that provide extensive opportunities for domestic interests to intervene in the policy process, as in the United States, it is much more difficult to develop consistent international policies, commit resources for international purposes, or accept onerous regulations required by international agreements when there is perceived conflict with domestic interests and priorities.

This strong national perspective—which quite naturally persists in the nation-state system, no matter how much erosion there has been around the edges of that perspective—is what makes problematic the prospect of truly extensive, uncoerced collective action that would seriously constrain national autonomy. As discussed in chapter 5, collective action on that scale may be necessary if a global issue (for example, global warming) is recognized as threatening the planet with catastrophe; but even in that case, the alternative—the assertion of hegemonic power by one or a few states—may be the more likely outcome.[17]

One particular form of international collective action, cooperation in science and technology, is obviously of special interest to this study. International cooperative programs, requiring varying degrees of joint planning, commitment of human and financial resources, and sharing of results, have been growing more common since World War II. The reasons are not hard to discern: important issues that cannot be tackled in one country alone, increased costs of research that put a premium on sharing, greater competence in science and technology throughout the world, stronger international mechanisms for coordination of planning and operations, and even political benefits that can be harvested from cooperative projects.[18] The number of cooperative projects is large, and it is certain to increase as subjects that *require* international research cooperation (such as the global environment) become more politically salient.[19]

For all these reasons, this period should be particularly propitious for more-extensive international scientific and technological cooperation; in fact, the record has been spotty, especially on ambitious, large-scale projects. There are many reasons, some of a purely practical nature that stem from the difficulty of meshing differing national budgetary cycles and systems of project selection.[20] The fact that in all countries the support of r/d is determined in a national policy process is the source of many of those practical problems.

But perhaps the most important reason is the lack of the kind of effective leadership required for equitable international collective action. The United States played a responsible leadership role during the period after World War II, when it was the dominant technological power. Now that dominance is gradually fading, as the natural result of the growth of European and Japanese capabilities.[21] The United States, though the strongest nation overall in science and technology, is nevertheless only one among equals in any given field, sometimes ahead of others, sometimes behind.

The United States has found it difficult to reflect this diminished technological dominance adequately in its policies toward cooperation. Several large cooperative programs have been undertaken with the implicit assumption in the U.S. government that the United States would determine the program's scale and objectives on its own and others would join in almost as subcontractors. Other nations were too often treated cavalierly, as an agreed project was restructured without consultation or even canceled without notice. The country has as a result earned an uncomfortable reputation as an unreliable partner in scientific cooperation.[22]

A quite different attitude is required for collective action to be successful among independent nations. There is a need for a leader to initiate cooperation, but there must then be genuine interaction, starting with the determination of objectives and initial planning and proceeding through the many steps of implementation. And there must be the certainty that commitments will be fulfilled. Those lessons may be difficult to accept for a nation that has been used to dominating programs with money and knowledge, for a nation whose policy process gives great weight to domestic interests and makes fulfillment of long-term commitments quite uncertain. But if those lessons are not somehow reflected in improvements in the policy process, international cooperation involving the United States will always be flawed. That would be a great loss to the United States and to the international community, in view of the United States' competence and the great size of its scientific and technological enterprises. Increased international cooperation in science and technology is an inevitable need for the future; that cooperation can come about efficiently and productively, or be subject to costly, and perhaps dangerous, impediments. The United States will have the largest say in which pattern prevails.

International Organizations

In the course of this study, we have seen the frequent need for international institutions that are able to cope with the international consequences of technological change or are necessary to realize the benefits of technology. The expansion in the number and role of international governmental organizations (IGOs) of both limited and universal membership, in this century and particularly in the last several decades, reflects this need.[23]

Permanent IGOs began to be established in the nineteenth century, and they have grown rapidly in number and scope in parallel with the growth of interdependence. Harold Jacobson lists 621 IGOs in 1980 (there were at least ten times that number of international nongovernmental organizations [NGOs], not including multinational corporations).[24] The large majority were created to carry out specific functions, often strongly technology-related—such as regulation of the radio-frequency spectrum (International Telecommunication Union), control and eradication of disease (World Health Organization), control of atomic energy (International Atomic Energy Agency), or regulation of commercial aircraft (International Civil Aviation Organization). Most of the major universal-member organizations are now specialized agencies of the United Nations. All such organizations, whether narrow in purpose or designed to serve larger political, security, or economic goals, bear some direct or indirect relation to science and technology, just as does international politics as a whole.[25]

For a time, many saw the organizations as precursors of an important change in the international political system. It was thought that the delegation of authority and power from nation-states to international organizations in order to carry out "essential" international functions would, over time, lead to erosion of the independent power of nation-states and to their gradual integration in a larger political entity.[26]

A considerable measure of such delegation has taken place, but not of powers whose delegation would seriously derogate from the independent authority of states. With the important exception of Western Europe, little political integration has taken place; nations have remained jealous of their prerogatives—even more so because they feel some have been diminished—and have tended on the whole to accept internationally–based regulation and management only when it was clearly beneficial and when vital national interests were not at stake. National policies toward IGOs are conditioned by the same factors as are policies toward international cooperation in general, making them hostage to domestic interests in nationally based policy processes. When IGOs serve or benefit domestic interests, official attitudes are supportive. When they interfere with or complicate achievement of domestic goals, official attitudes tend to be hostile.

It is nevertheless true that the role and functions of IGOs are a considerable factor in the gradually increasing constraints on national autonomy that all states have experienced as interdependence has intensified. It is also true that any organization, no matter how circumscribed in its charter or oversight, develops some measure of autonomy and independence over time. That applies equally to IGOs; they are in many ways not the totally servile eunuchs of nation-states they are often assumed to be.[27] And they go through a process of learning and adaptation as their framework evolves and as their primary purposes gradually change.[28]

Not only do IGOs perform essential functions, but those with a largely technical purpose are in fact quite effective in their operation, more effective than is often assumed.[29] They have to be. The requirements for managing an increasingly integrated world could not have been met without them, and their existence and performance imply that nations recognize they cannot achieve their national goals, domestic or international, without an international structure that includes a growing number of reasonably autonomous and competent multinational organizations.

An obvious question is whether the growing need for effective collective action, a requirement that accompanies technological change, will result in steadily increased delegation of authority, perhaps leading to a more substantial evolution in the role of IGOs sometime in the future. In 1946, the serious attempt to internationalize control of the atom foundered on the rock of big-power politics, even in that time of idealistic and, as it turned out, unrealistic hope for the future authoritative role of the UN.[30] Perhaps tomorrow, threats of catastrophe, whether ecological or economic, will make it imperative and acceptable to delegate substantial economic and political authority to international organizations.

It is perhaps more likely that in the face of a crisis, one or a small number of states, rather than accept the growth of authority of IGOs at the expense of the authority of nation-states, will take matters in their own hands and carry out necessary actions through more-traditional means of exerting power (as discussed in chapter 5).

A possible harbinger of changing attitudes toward international institutions may be the use of the UN to mount a genuine collective response to aggression when Iraq invaded Kuwait in 1990. That response may have been possible only because of the clear-cut violation of international norms and the absence of old cold war rivalries. The more important precedent for the long run may well be the unanimous Security Council authorization of an intrusive monitoring and inspection role for the UN and the IAEA in Iraq in the wake of the Gulf War, in the search for weapons of mass destruction.[31] That action represents an unprecedented mandate for interference in normal sovereign prerogatives, agreed to in extraordinary circumstances but of potentially great importance for the future role of the UN.

The development of the European Community represents a rather different example of possibly changing attitudes. That assemblage of reasonably like-minded states of roughly comparable standards of living has been making tortuous but steady progress toward aggregation of responsibility in the European Commission, with substantial ceding of authority to that body.[32] Impressive progress has been made, demonstrating that independent nations can recognize the limitations of their nationalism and the advantages of pooling their sovereign powers. But the reunification of Germany, the collapse of the military threat from the East, and the eagerness of additional nations to join represent new conditions for the Community that are likely to slow the momentum toward genuine unification. The political dimensions, in particular, are uncertain (the Community did not perform well as a unit in the political run-up to the Gulf War).[33]

The experience of the Community does show that under certain conditions, mature nations with roughly similar economic and political conditions are willing to sacrifice major elements of their autonomy when the benefits are seen as substantial and the alternatives sufficiently undesirable. Those certain conditions may, however, be unique to Western Europe.

It should be noted that the expansion of the number of IGOs is far outstripped by the increase in the number of NGOs operating in an international environment.[34] It is not surprising, for the growth of cross-border relationships that is a part of interdependence naturally includes increased international collaboration among domestically-based interest groups. NGOs have become an important part of the international scene, in no small measure because of their effective use of the fax machine, the media, and other information technologies. In some cases, they have come to play quite prominent and influential roles in the evolution of policy.[35] Many are organized around environmental issues, and they were expected to be an important presence at the 1992 UN Conference on Environment and Development in Brazil.

The extent of the policy influence of NGOs is not uniform among organizations or subjects, nor can it be estimated responsibly without extensive research. Their influence, however, is likely to grow over time, as they accumulate more members, prove increasingly adept at mobilizing public opinion, and exert pressure on more governments. Over time, NGOs may well become a major factor in the evolution of the international political system as a "third" force that leads nations to turn more readily to international bodies on transnational and global issues, and to accept collective action.

Important questions can be asked about their role, however. NGOs, being private, are not representative organizations; the participants, often the self-appointed leadership, set their own agenda. They are organized typically around narrow issues, sometimes only a single issue that allows little room for interaction and compromise with the agendas of other NGOs or governments. And their usually narrow and committed advocacy role frequently raises concerns about their sometimes-biased use of science and scientific information.

Perhaps the largest impact of NGOs will be in the global propagation of a common set of values. This prospect worries developing countries, since many believe NGOs tend to be dominated by the values of industrial countries. The result is that what NGOs are advocating, with increasing international acceptance, may in fact be considered inimical to the long-run interests of developing countries.

Some Additional Issues

These governance issues arising from technological change drawn from a much larger set, are the ones most likely to have, over time, the greatest effect on the evolution of international politics. Some other issues that have emerged in this study, however, are of more than passing importance in their implications for governance.

Uncertainty

Uncertainty figures often and importantly in almost all of the issues we have explored and is, of course, a necessary part of all policy processes. The quantitative nature of scientific and technological aspects of policy, however, at times conveys a sense of precision—which, as we have seen, is often wrong or misleading. Moreover, as more global-scale issues emerge, the uncertainty that must be confronted in policy processes is likely to grow larger, more complex, and more consequential, greatly complicating the development of collective international action.

Excessive Information

The increasing generation and flow of information worldwide creates major problems for policymakers. Even the application of technology to make more manageable the information that must be considered in the formulation of policy does not overcome the problem of information glut, for technology cannot separate in advance the important from the merely relevant without sacrificing to the programmer the decision as to what *is* and *will be* important.[36]

There can never be a completely satisfactory solution to this problem, because the generation of relevant information will increase as global issues proliferate and as technology becomes better able to measure all relevant variables of an issue.[37] Handling and assimilation of information will, as a result, continue to be a characteristic and growing problem for all governments.

Process for North-South Transfer of Resources

Several issues discussed in the earlier chapters have important implications for North-South relations, but more than any other, global warming—should it become a clear-cut threat—will affect those relations by posing a need for substantial transfer of technology and resources from industrialized to developing nations (as elaborated in chapter 5). Such a transfer might have to be on a far larger scale than current transfers from industrialized to developing countries, raising major questions of how it could be accomplished by governments that have found the process to be exceedingly difficult even on today's much smaller scale.

Recognizing (and accepting) the need may be the first and most important step. Beyond that, new patterns of relationships will have to be forged that will take North-South transfers out of the "foreign-aid" framework. Many ideas are extant—for example, tradeable permits that would allow developing countries to generate resources by selling emission rights; or the establishment of policies in industrial countries that would recognize the global nature of greenhouse-gas concentrations and that would empower domestic energy or environment ministries to carry out emission-reduction activities in whatever country they are most cost-effective.[38]

Dependence on Large Technological Systems

The increasing dependence of society on large technological systems, discussed in chapter 4, affects matters of governance in many ways. One of the most challenging problems is building the capacity to respond to the inevitable crises that will arise from system disruption.

This is a generic problem for governments; it is relevant to large, complex systems in general and particularly to those that are heavily technological in nature (e.g., computer information systems, electricity-distribution grids, and telephone systems). Measures to protect against system breakdown are necessary, but perfect protection is not possible (short of abandoning the technology altogether), and breakdowns are certain to occur. The policy processes of states, singly or in consort, must be prepared for breakdown and ready to deal with its effects, including longer-term consequences, without knowing in advance how disruption will occur or what its dimensions will be. That is not a simple challenge for a state coping with an internal threat; it is much more difficult in an international environment among sovereign states. It is possible that eventually, if the crises resulting from system breakdowns are sufficiently costly, new international mechanisms with substantial emergency authority (at least) over activities within nations may prove to be necessary and politically acceptable.

Public Involvement and Political Fragmentation

One of the major consequences of the greater public availability of information that accompanies the spread of information technologies is increased public involvement in policy processes. Clearly, the extent and significance of this involvement will vary from country to country and from one political system to another. The effects in Eastern Europe and the Soviet Union, as discussed in chapter 4, proved to be of critical political importance in leading to change in the governmental structures of those countries. In most Western countries, the greater availability of information and easier access to it have served to support, to differing degrees, increased public participation in policy processes. This is generally applauded as a contribution to democratic goals and as an important step forward in the functioning of Western democracies.

On the other hand, that same increase in information availability has made it easier to organize interest groups around a single issue to influence policy processes. In the United States in particular, the result has been an increased fragmentation of politics and more-frequent gridlock in the policy process as the multiplication of interest groups has made trade-offs and compromises harder to achieve.

The workings of policy processes are affected not only by greater public participation *in* those processes, but also by the increasing ability of groups forming within or among nations to bypass established lines of authority and operate on their own. The proliferation of subnational and supranational actors often serves to constrain the autonomy of governments; in turn, that constraint on governments serves to encourage and empower even larger roles for those nonstate actors.[39] These structural developments of politics involving greater participation and fragmentation have multiple causes; technological change is but one factor. New technologies, however, particularly information technologies, are important contributors and are likely to continue to stimulate further change in the structure of politics and will continue to do so.

In Sum

The many effects of technological change on the nature and problems of governance occur only gradually, which would seem to allow time for policy processes to adjust. However, as with all large systems—and government policy processes are large systems—changes come very slowly, with long time lags between adjustments, and often only in the context of a crisis of some sort that grows out of the mismatch between the need for new policies and the actual decision process to bring them about. The effects of technological change make timely improvement of governmental processes even more difficult than it has been, in part as a result of the rapidity of change and in part because it is not possible to overcome completely the new problems introduced through technology (e.g., modification of the time frame for decision making, dependence on machine processing of information for urgent decisions, unequal access to the technological elements of issues, or unpredictable breakdown of essential technological systems).

The recognition that governance is never as effective as it might or should be is not novel. The problems introduced by technological change, however, increase that gap and make it harder to achieve the level of effectiveness that a more complex, technological world now demands. The technology-related problems of governance thus become an added factor in their own right in the evolution of international affairs.

PART THREE

Seven

Conclusions and Observations

IN A WORLD that is less than fifty years from the discovery of the structure of DNA and the invention of the transistor, only a few years longer from the key experiment that demonstrated atomic fission, and only a little more than a century from the invention of the telephone—out of some nine thousand years of recorded human history and some three hundred thousand years since the appearance of *homo sapiens*—it may well be a quixotic exercise to attempt to describe with confidence the effects of technological change on any contemporary aspect of human affairs. The brief time in which technological development has taken place in the totality of human experience not only raises questions about any generalities that might be made, but also clouds judgment, encouraging grand statements while impairing the ability to be skeptical about the deeper significance of the material changes that are so evident all around us.

The difficulties apply strikingly to international affairs, for it takes a strong act of will to preserve objectivity about the political significance of change while recognizing how much and how fast the world scene has in fact been altered because of science and technology. Objectivity is not made easier by the enormous problems in measuring the extent of change and in defining when a change becomes something more than incremental. Moreover, the increasingly porous boundary between domestic and international affairs—one of many developments in which science and technology have played a major part—means that the whole human condition must be taken into account when assessing international affairs, to an extent that was not necessary—or was at least considerably less necessary—in the past.

These dangers to analytical sobriety notwithstanding, the study presented in these pages did attempt to look hard and critically at that dynamic relationship between technological change and international affairs, with as steady and nonrhetorical a hand as possible. A review of the most significant conclusions of this study will make possible an integrated overview of the entire, occasionally all-too-amorphous, subject.

Some Summary Conclusions and Generalizations

Our initial purposes were to ask how scientific and technological change has influenced the evolution of international affairs, and whether the characteristics of the modern scientific and technological systems and their close interaction with international affairs have influenced that evolution in particular directions and might be expected to continue to do so in the future. Along the way, we considered whether some elements of that evolution that have already taken place, or are in prospect, could be held to be fundamental changes that may have caused, or may presage, more-appreciable changes in the structure of the international system.

The study has shown many of the dimensions of the quite-substantial evolution in international affairs that has accompanied technological change, and how characteristic technological outcomes have influenced the direction of that evolution in particular ways. But the study has also shown that that evolution has been incremental in nature, posing only limited challenges to traditional assumptions and concepts. Only a few, very few elements of international affairs can be said to have been fundamentally changed. For the others, even substantial evolution has not altered basic concepts, or at least has left the extent of their alteration indeterminate or ambiguous.

A summary that is focused on customary themes in international affairs will both highlight the most significant elements of change and enable us to see more clearly its extent and nature.

Sovereignty: Autonomy and Authority in Economic and Political Affairs

One of the central organizing elements of international affairs, cited often as the element on which science and technology have had the largest impact, is the principle of sovereignty, "the constitutive principle of the existing international system."[1] As noted in chapter 1, the term has various technical and popular meanings; it can be used to be entirely coincident with the meaning of statehood, or it can mean, less precisely but more commonly, the authority and autonomy of governments in ordering a nations' affairs. The latter meaning is by far the most frequent, even in theoretical discourse.

Many commentators describe the increase of economic interdependence, or the rise of multinational corporations, or the growing role of international institutions, or intensified worldwide communications, or the threat of nuclear attack or proliferation—all of which were greatly stimulated or made possible by technological change—as the primary cause or causes of the erosion of national sovereignty. By that is meant that the emergence of those phenomena on the world scene has resulted in growing limitations and constraints on the authority and autonomy of national governments. There is no doubt of the reality of this conclusion; there is today a large and expanding sector of national and international activities not under the direct control of governments, nor accountable to them, that impinge on the authority of governments and constrain to varying degrees their freedom of action or ability to order events. This is arguably the most significant aspect of the evolution in international affairs that has accompanied technological change.

Sovereignty in its narrower conception, equated to the basic idea of statehood, is by definition not much affected by the rise of alternate sources of power and influence outside the direct control of governments. The realist school of international relations, for example, would argue that those alternate sources of power are dependent on the fundamental power of the nation-state, not separate from it.[2] In this view, the world remains an anarchic system of independent, sovereign states in which states cooperate for the purpose of serving the national interest—or are forced to cooperate by more-powerful governments—and allow the growth of interdependence for the same reasons. Constraints on the authority and autonomy of government are not seen as fundamental developments that challenge the underlying fact of independent statehood.[3]

Under either concept of sovereignty, governments must now recognize a host of new or enhanced conditions in the making of policy that blur the lines between domestic and international affairs, that increase the relevance of other countries' actions to domestic interests and goals, that cause domestic policies to have important repercussions on the fortunes of other nations, that raise the costs of the pursuit of autarky, and that by implication create an inescapable requirement for extensive cooperation and interaction with other governments and societies. All of these conditions have been present since interdependence became an important aspect of intercourse among nations; but they have been greatly intensified as technological change has expanded the incentives, the opportunities, and the requirements for transnational linkages.

Does this genuine evolution in international affairs constitute a fundamental change in the international system? Is the change in the autonomy and authority of the state large enough to be legitimately considered a system change? Though it is necessarily a matter of definition, of whether there are thresholds above which incremental change becomes a change in kind, some clearly believe that such a point has been passed; as James Rosenau asks with more than a little impatience, "How much evolution must there be before it can properly be [viewed] as basic change?"[4] In our study we have not minimized the importance of the substantial evolution in elements of the system that has taken place. But the overall conclusion is that the proposition of basic systemic change remains unconvincing, given the ambiguous effects of change and the opportunities that continuing technological advance offers for governments to bypass or resist constraints on their autonomy or authority. This conclusion holds, as discussed later in the chapter, whether the realist or liberal school of international relations is accepted as the dominant theoretical paradigm that describes the international system.

In the international economy, for example (security issues are discussed below), even the dramatic developments of global financial markets and multinational corporations among market economies are, as we saw, less independent of national policy than at first appears. Governments retain many levers of influence; and technological change offers, as one of its more important effects, new options for policy that enhance the possibilities for governmental intervention and control.

Our conclusion that technological change has had important but limited effects on governmental autonomy and authority may be valid as a general proposition, but we have seen two significant exceptions: one that has to do with constraints on the economic organization of a state in a competitive technological environment; and another, equally important but somewhat less certain, concerning limitations on authoritarian political control by governments.

The first exception with profound political as well as economic implications, is the practical inability of a centralized command economy to be competitive with market economies in high-technology innovation and products. Command economies, as we observed in chapter 4, cannot satisfy the conditions of sustaining scientific and technological enterprises able to bring new technology to the market at competitive price and quality. Economies, of course, are never pure examples of one or another form of organization, but those that approximate a command structure are certain to be deficient in innovative capacity relative to those organized dominantly along market lines. The command economies of Eastern Europe were prime examples; their inherent weakness in the technological capacity necessary for growth and competition proved to be a key factor in their ultimate collapse. A nation might elect to retain a command structure and forgo technological competition and its fruits; or it might lack the resources to compete in technology, whatever its economic organization. But if it aspires to have an innovative economy able to compete with market economies, a command structure is not an option.

The other exception, closely related, is the impact on the political authority of the state that results from the introduction and spread of new technologies, especially information technologies, that foster decentralization of political power and limit the ability of a government to exclude unwanted ideas and information. These technologies tend to support conditions that lead to pluralistic and democratic forms of government, and correspondingly tend to undermine authoritarian political structures. Authoritarian forms of government cannot be said to have become unrealizable, however, for it is possible to use these same technologies to stifle movement of information and decentralization of power, and political power can be maintained through raw use of force; several historical and current examples make the point.

Nevertheless, there does appear to be, for reasons that we examined earlier, a clear bias in the political effects that follow the introduction of such new technologies, a bias that favors the liberalization and decentralization of power. And this bias appears to be reinforced as technological development continues. Over time, therefore, new technologies will, on balance, make it more difficult for authoritarian forms of government to maintain power; the result will be fewer such regimes and shorter lifetimes for them when they do appear. If this proposition continues to be valid in the future, it must also be considered to be a substantial change in a central element of world politics.

Competition and Dependency

States strive for economic gain and influence in an interdependent world. They are in competition with other states and at the same time are increasingly dependent on them. This is not a new development, but the effects of technological change have altered in several ways the determinants of these economic relationships. A few key elements stand out in their larger significance for international affairs.

COMPARATIVE ADVANTAGE

Comparative advantage is a central concept in international trade that describes the relative economic position of nations. It explains the flow of trade in a positive-sum free-trade system in which nations trade those goods and services in which they have a price or quality advantage for those in which others have an abilities.

The source of a nation's comparative advantage was traditionally seen as its endowment in factors such as natural resources, land, and capital. As technology-intensive trade has become a larger proportion of international trade and has become more significant in economic competition, it has led to an important change in the sources of comparative advantage. Success in technology-intensive trade is determined not by a state's endowment in natural resources or productive land, but rather by the skills of its people, the quality of its science and technology, its capacity for technological innovation, and the effectiveness of policy and management. In short, comparative advantage in technological goods and services is "created" by a society through its policies and human abilities.

One implication of this change is that issues of domestic policy previously considered to be relatively independent of trade matters (e.g., r/d funding, education policy, government procurement, antitrust policy, intellectual-property rights, and even financial and industrial structures) are now important determinants of a nation's competitive performance.

A corollary is that these domestic matters are now inevitably embroiled in international trade negotiations; nations seek agreement on rules about matters that reach deeply into a country's internal policies, and even to its economic structure and culture. Whether agreements can be reached to reconcile often quite basic differences among nations in these matters is not at all clear; if they are possible, the compromises necessary for some nations could well require significant modifications of internal structure as well as policies.

These are important changes, with some potentially large effects as nations come to grips with the subtle and sensitive issues that tie international trade so closely to domestic policies. The changes illustrate well the loss of a real distinction between domestic and international affairs, but they do not appear to signify a larger, systemic change. Different players may dominate the international trading system, but the system will stay basically the same.

EXPANDING OPTIONS

One of the most influential capabilities offered by science and technology is the possibility to expand, through targeted r/d, the options that are available to serve national and industrial policy objectives. This is seen, for example, in the growing ability to target r/d to find means to bypass an undesirable resource dependency (through increased efficiency of use or through development of substitutes or alternative technologies), or to alter a dependence on low-cost labor in foreign production (by automating a production process). This new degree of flexibility has changed the nature of many existing dependencies, especially those between the developing and industrial nations, by modifying (usually reducing) the comparative advantage of countries that have only raw materials or lower-cost factors of production to trade.

The ability to mold technological outputs to counter any dependency is not by any means total or immediate, dependence on oil will not be quickly or easily altered, nor can all factor-cost advantages be summarily neutralized. However, it is a steadily growing capability that expands as knowledge expands; and it is a potential and fertile option for policy, one that implicitly constitutes a challenge to the continued existence of most dependencies. It serves in this way to change the content and significance of many of the traditional geopolitical elements of international politics.

SCIENCE- AND TECHNOLOGY-BASED DEPENDENCIES

Among industrial countries, the rapid improvement in the scientific and technological competence of European countries and Japan has ended the overpowering postwar dominance of the United States in those fields. Though the United States remains the nation with the broadest commitment to all fields of science and technology, the pattern now is one of roughly comparable capabilities in science and technology across national borders, with some variation in capabilities for the innovation necessary to turn laboratory advances into commercial products. As a consequence, a nation's performance in science and technology, vital to its competitive success, is now dependent not only on its indigenous scientific and technological capabilities, but also on its ability to stay abreast of progress in other advanced countries and to use that progress effectively in its own research and innovative activities.

In this situation, a new form of dependency could be emerging in which a strategic technology that has positive externalities across a range of industrial sectors comes to be dominated by one country only, to the exclusion of others. In the extreme case, such a technology could be considered "critical" if it were essential not only for a substantial number of direct applications, but also for the ability to design products that are competitive in price and quality in a wide range of other high-technology areas. If one country attained such a dominant position, it would have an economic (and potentially political) advantage that could be exploited for a variety of national ends.

Japan's lead in the manufacture of semiconductor memory chips, which are the essential components of virtually all electronic products, is the case most often cited as an example of such a critical technological dependency; as we saw in chapter 4, however, the extent and permanence of that Japanese dominance remains uncertain and controversial. If the dominance turns out to be valid and long-lasting, the political effects could lead to a substantial reshuffling of economic winners and losers among nations in coming decades, though without necessarily causing any more-fundamental changes in the structure of the international political system.

Another new form of technology-based dependency is a result of the development of global technological systems. Applications of space technology in particular, such as communications, resource sensing, and weather forecasting, have led to global systems on which countries have (or will) come to depend for the provision of essential services, services that gradually become integral parts of national economies. These systems are owned and operated by national governments, occasionally by international organizations; in either case, the owners become responsible for continuity in the operation and quality of the service on which national economies have come to depend.

Nationally owned systems are vulnerable to decisions made in the context of domestic politics and economics; continuity may therefore be determined by domestic needs, not by the needs of other nations using the system. Over time, if nationally based systems perform poorly or fail, or if their continuity appears threatened, there will be interest in the transfer of ownership to international organizations where many nations would have a voice. The migration of large technological systems to the control of multinational bodies, if it were to take place, would contribute substantially to the gradual growth of the responsibilities of multilateral organizations in performing functions that are typically the province of national governments today. It is more likely, however, that poor or parochial performance of a nationally operated global system, even a costly one, would lead other nations to develop competitive systems, rather than encourage a move to an internationally operated system.

The ability to present dramatic information about technological catastrophes instantaneously around the world (and the inability to prevent their becoming known) creates another and unusual form of technology-based dependence. The viability of a well-regulated and competently operated nuclear industry in one nation, for example, is in an important sense hostage to the possibility of nuclear accidents in countries in which the industry may be poorly managed. If nuclear power becomes an essential technology in widespread use (perhaps in order to minimize global warming), this issue will have to be dealt with, for it places the viability of the industry at the mercy of its weakest national component; nations now heavily dependent on nuclear power already must face the issue. Presumably, some new, fail-safe reactor design would have to be adopted, with international regulation and oversight that could provide greater assurance of the quality and safety of design and operation. It would be another example of the movement of formerly national responsibilities to a multilateral framework.

One of the more troubling and politically sensitive forms of technological dependency is that between industrial and developing countries. Developing nations, even if effective in improving their indigenous capability in science and technology, continue to depend for their economic growth on the ability to tap knowledge being generated in the technologically advanced countries. The dependence results in often-contentious North-South disagreement over requirements for effective transfer of technology, flow of technological resources, intellectual-property rights, protectionism, "brain drain," technological terms of trade, and similar matters. Some of these issues have a chicken-and-egg quality; for example, successful technology transfer requires, *inter alia*, an existing capability in science and technology to receive the transfer and to be able to adopt it effectively.

The situation is made more costly for developing countries by the increasing options science and technology present to industrialized countries for avoiding dependence on resources, location, or other comparative advantages the developing countries may have. Moreover, the national basis of decision making for r/d means that developing countries are hostage to the results of r/d decisions in which they had no part. That is by no means a unique situation, but it can be particularly damaging to the interests of countries likely to be vulnerable to a very few decisions made elsewhere that will have large effects on their terms of trade.

This technological dependence of third-world countries on those of the industrial world is likely to be long-lasting, though it is highly variable among countries, with an appreciable number of formerly poor nations moving toward industrialized status. It is not a new phenomenon; and it will not lead to larger system changes, unless the requirements of dealing with planetary-scale dangers force industrialized countries to adopt entirely new approaches to meeting the goals of the technologically poorer countries.

Finally, the importance of science and technology in economic growth and competition means that the economic health of a nation—one of the determinants of its comparative advantage in international commerce—is now dependent on the structure, quality, and strength of its indigenous scientific and technological communities. This is a new and important element in dependency but, as noted earlier, not one likely to lead to change in the international system (though it could certainly change the identities of the key players).

LARGE TECHNOLOGICAL SYSTEMS

A distinguishing characteristic of modern society is the evolution of large technology-based systems, an evolution stimulated by the characteristics of technology that put a premium on interconnectedness and economies of scale. Large systems have become crucial elements of national economies and military structures, and are increasingly international—often global—in their operation and significance. They are one of the hallmarks of growing global integration and are a pivotal element of interdependence.

We have seen in both the security and economic domains that dependence on large technological systems can create dangerous instabilities. This is especially so as the systems become more tightly coupled, requiring real-time information and immediate responsiveness. In the security sector, the disruption or malfunction of command and control systems for nuclear forces carries the danger of precipitating nuclear conflict. In the economic sphere, system disruption can entail large costs and, in some situations, significant economic and political disruption.

The vulnerability of large systems can be reduced, and technology can help in that task. But there is a trade-off between performance and vulnerability; the risk of breakdown or malfunction can be removed entirely only by effective abandonment of a technology.

The problem of vulnerability is made more serious by the fact that the operation of some technological systems often precludes meaningful human intervention in times of crisis or unusual volatility. The command and control systems for nuclear forces again provides an apt illustration; the scant time available for decision, and the importance of the analysis and interpretation of very large amounts of ambiguous information, mandate heavy reliance on computer analysis to determine operational orders. Similarly, global financial markets must rely on preprogrammed computer operation, with little or no human intervention possible, during times of large volatility in the market.

Large technological systems thus lead to an intense form of dependency, both in terms of the scale of the effects that could result from system disruption and in terms of the degree to which society has been willing to accept constrained human participation and the corresponding dependence on preprogrammed machine decisions for crucial operations within those systems. The dangers inherent in this dependency, particularly in the economic sphere, are increasingly serious and are likely to be destabilizing factors in international affairs. They may eventually require new levels of international cooperation in the design and operation of systems, in the measures taken to protect them from intentional disruption (by terrorists, for example), and in the steps needed to respond to the crises that will follow inevitable breakdowns.

Those steps are not likely to receive the attention they deserve, however, until a crisis of some kind forcefully demonstrates the dangers of this dependency and the need for international measures to minimize their costs.

Military Force and Nuclear Weapons

In the security area, the most striking change has been the introduction of nuclear weapons and, in particular, their coupling with intercontinental ballistic missiles. The result, however, has been the creation of forces that can fulfill the traditional function of protecting the state and its citizens against a nuclear-armed opponent only as a deterrent, not in actual warfare; and there is no realistic expectation that developments in technology will change that situation. Nuclear weapons may be acquired to deter their use by others, to attempt to threaten non-nuclear states, or for political and prestige purposes, but they no longer can be used in anger against a similarly armed rival for the purpose of ensuring the survival of the state. In fact, they also turn out in practice not to be an effective threat against non-nuclear states, though they might be seen to be effective (as a deterrent only) for a state whose potential non-nuclear adversaries heavily outnumber it in conventional forces.

Since "war has always been, if not the essence of international relations, at least their distinguishing characteristic," this inutility of nuclear weapons in a war-fighting role is by any calculation a substantial change in international relations in general, and in the meaning of military power in particular. It has had important effects on the postwar organization and functioning of the international system, though not in the direction or to the extent many anticipated in 1945.[5]

To a surprising degree, this change in concept has been assimilated into international politics with a minimum of deeper change. Forces of unprecedented size and power have been built by two countries who saw themselves in political and ideological competition. For this and related reasons, a bipolar political structure previously unseen in the modern era dominated the postwar world until 1989, but the massive forces could never be used in conflict between the two superpowers. However, the use of force has clearly not been banished from the scene; more or less traditional competition for power and national advantage has proceeded among the now very much larger number of nations, many of which have come into being since 1945.

The 1990s will witness a reworking of the structure of forces among nuclear nations and their allies, as a result of the demise of the Warsaw Pact and the Soviet Union. NATO will be massively altered by the change, though the ultimate configuration of forces in Europe and the United States is not yet clear. But nuclear weapons continue to exist, and the knowledge necessary to produce them cannot be eliminated; in fact, it can only spread more widely. Nuclear weapons will remain essentially impotent for war-fighting purposes, but that does not mean there will be none in the arsenals of the major powers or in the forces of other nations determined to have them. They will continue to serve purposes of deterrence, threat, and prestige.

One of the ironies resulting from nuclear-weapons technology is that a total ban on nuclear weapons is not a sensible goal; it could give a new entrant an immediate and very large advantage in power against which there would be no adequate deterrent.

Evolution and Diffusion of Conventional Weapons and Power

There are likely to be profound effects on the conduct and meaning of conventional warfare as the outcomes of r/d lead to a steady upgrading of conventional weapons—upgrading that will include, to give but a few examples, increased firepower in smaller and lighter weapons packages that can be carried and fired by one person and that are affordable to poorer countries; longer-range delivery systems that are faster and more accurate, allowing effective application of force at a distance; improved ability to destroy high-value targets, such as ships, at relatively low cost; and much-improved real-time surveillance capability.

Evolution in armaments is inevitable as both commercial and military resources continue to be devoted to r/d in a system that rewards maximum change and innovation. The equally inevitable spread of technical knowledge means that in time, even in the unlikely event of more-effective control of the arms trade, a substantially larger number of countries will be able not only to purchase, but also to produce, what are now considered to be advanced weapons. The political and economic incentives for arms producers to sell weapons continue to be strong, supported on the demand side by the assumption of the value of technological superiority—an assumption most recently strengthened by the evidence of the 1991 Gulf War. Moreover, the very rapidity of technological change in weapons encourages the sale of those weapons made obsolete by new designs. Even the otherwise-gratifying end of the cold war contributes both surplus weapons and, more significantly, surplus arms-production capacity to the market.

As a result of all these developments, the military reach and power available to small states will grow, enlarging the area of involvement, the destructiveness, and the danger of escalation of local conflict. Instabilities will likely flow from local arms races fueled by the appetite for advanced weapons, and there is the danger that a transient superiority in weapons might tempt preemptive attack. In time, and perhaps not far off, the diffusion of advanced conventional capability will make potential conflicts between a major power and a small nation less one-sided in potential for damage and correspondingly more dangerous.

These are important changes that carry the threat of a more perilous and unstable world in the future, but they do not necessarily challenge traditional notions of the purpose and use of force. Perhaps if these developments were coupled with the proliferation of nuclear, chemical, and biological weapons of particularly destructive potential, there would be such a challenge. Countries that acquired such arsenals capable of massive destruction and that were faced by similarly armed opponents would presumably find that the imbalance between the costs of the use of force and any potential gains would be as potent a deterrent for them as it is for confrontation between nuclear-armed nations today. But the risks of a breakdown of deterrence would be much higher, given the wide availability of weapons of mass destruction, the greater vulnerability of some nations than others to attack, and the varying level of knowledge and rationality among national leaders—in addition to all the associated problems of control, misinterpretation, and domestic turbulence. In those circumstances, there could be little confidence in the view that acquisition of weapons of mass destruction in effect ruled out the possibility of their use as a means to achieve national goals.

More generally, if the potential danger and destructiveness of local war reached a sufficient pitch, as it well may, it is possible that nations would take more seriously the goal of an international condominium that genuinely attempted to control the diffusion and use of all weapons. Such a development would be a major change in world affairs if it proved to be effective in the control of weapons and warfare; it certainly would imply a fundamental change in the geopolitical meaning of force, as well as in many other aspects of international affairs.

There are some encouraging new developments that grew, typically, out of adversity. The spreading capability for the production of ballistic missiles was sufficiently worrisome to Western nations that it led to the creation in 1987 of the Missile Technology Control Regime (discussed in chapter 3) to attempt to control, or at least delay, further diffusion of relevant technologies. Similarly, the shock after the Gulf War when it was discovered that Iraq had been successfully pursuing, without detection by the IAEA, several paths for acquiring the material and equipment necessary to build nuclear weapons is proving to be a powerful spur to the strengthening of the IAEA's capability for detection of clandestine nuclear programs. Both of these developments will mean a significant increase in the security role and authority of international organizations.

The end of the cold war gave rise to the hope that the growing importance of economic as opposed to military affairs in international relations signals the beginning of a major shift in international politics away from conflict as a policy option. That shift may well have begun for the industrialized countries; certainly, the nations of Western Europe, with such long histories of repeated conflict, have found it possible to create within a few decades an environment in which internecine warfare is all but unthinkable. Can such an environment be extended eventually to other parts of the world?

Despite these mildly hopeful signs, the evidence is not encouraging; conflict continues in the third world, the arms trade proceeds unabated, and nationalism and ethnic rivalries have reemerged as powerful sources of fragmentation and dispute within and among countries, including, most vividly, the independent states of the former Soviet Union and Yugoslavia. In that environment, the prospect remains high for continued diffusion of power and steady growth in the capabilities of the armaments available to third-world countries, and little modification is to be expected in traditional attitudes toward the use of force.

Other Geopolitical Factors

It has been evident throughout this study that the traditional factors thought to be significant in geopolitical assessments of a nation's international power and influence have been substantially altered as a result of advances in science and technology. There is no single list of such factors that is commonly agreed on; the more important ones—those concerned with military power, resource endowment, and sources of economic competitive strength—have already been summarized in this chapter. Three others deserve particular mention, however, for they have always been high on lists of geopolitical factors and have been importantly affected by technological change.

Geography

Geographical position is clearly of less significance than it was in the past, both in and outside the security domain. Radical changes in the technologies of transportation, communications, and weapons have undermined the security of distance, the option of autarky through physical isolation, and the economic advantages of location.

Geography is by no means completely irrelevant; it does, for example, affect how rapidly military forces can be transported where needed, the physical vulnerability of a nation to specific weapons systems, and, in economic affairs, the formation of regional blocs and special economic relationships. But it no longer carries the weight in world affairs that it once did.

Population Size and Skills

At one time, a large population was critical for mounting adequate military forces or for maintaining a robust economic base. It is no longer a primary determinant of the military power a nation can assemble, though it does retain a correlation with economic strength. The economic relevance is not only in one direction, however. In some nations, a large population contributes positively to economic potential; in others, a large and rapidly growing population is an economic drag, not an advantage.

The level of a population's skills is becoming the more important geopolitical measure. The necessity of technological capability to operate a modern society and to be competitive in a technology-intensive world has greatly increased the importance of a nation's ability to train a technically competent citizenry and to maintain a high-quality education system.

Quality of Governance

The capacity of governments to cope with rapidly evolving issues and relationships in a system more nearly geared to stasis is under severe challenge. The role of governments and the issues with which they must deal are changing, in good part under the influence of scientific and technological developments; moreover, the technical aspects of policy processes and the second-order consequences of technological change are placing new and unprecedented demands on structures created under, and responsive to, quite-different conditions. At the same time, the very basis of political processes is evolving as new technologies contribute to the decentralization of domestic politics and to the development of a more exposed form of "media" government. Effective governance can be aided by the use of new technologies; but, on balance, the difficulties (explored in detail in chapter 6) continue to grow.

In that more demanding environment, the differential capacity of states for effective policy-making—and particularly the capacity for productive use of scientific and technological resources, which are now so critical to international competitiveness—will be a major determinant of states' relative international position. This differential capacity is already in evidence; some nations, most notably Japan, appear better able to meet the new requirements. It is uncertain whether that situation will continue as Japan faces problems it has so far been able to ignore (for example, the need to improve its capability for basic research or the need to respond to demands from trading partners for change in domestic economic structures); major consequences, both economic and political, will ride on the relative quality of Japan's governing processes compared to those of other technological powers.

New Issues

One of the other striking elements of the evolution in international affairs that has resulted from advances in science and technology is the emergence of new or greatly modified issues with substantial political or economic significance. The list is lengthy, including atomic energy, space, the environment, information technologies, agriculture, energy, and others. Some of these, particularly information technologies and the environment, raise issues that have the potential to alter the international system in substantial respects; most are clearly only evolutionary in nature—posing important questions for governments, stimulating new international regimes, becoming foci for competition and dispute, but for the most part adding to the agenda in relatively familiar ways.

One set of new issues was singled out in chapter 5, as potentially of much larger significance for international affairs in coming years. These are issues of planetary scale that may pose threats to all people and all countries, issues we have called global dangers.

Climate change as a result of global warming is the best example and most publicized of such dangers. Though it is not usually described this way, an increase in the mean temperature of the atmosphere, if it occurs, would be a result of the interaction of two large global systems: the natural ecosystem and the human-built socioeconomic system. The latter now is large enough that for the first time in the planet's history, the by-products of normal human activities—emissions from fuel combustion and other greenhouse gases—can alter the entire ecosystem, with effects that are at present uncertain. Some climate change is likely to take place, and there is a reasonable probability of change substantial enough to have significant economic and social effects. There is a small but non-zero possibility of climate change that would be catastrophic in its effects if the rate of increase of greenhouse gases stresses the ecosystem sufficiently to cause nonlinear, unpredictable shifts of system elements, such as ocean currents. Though the large inertia of the two systems makes overall change in either system slow to bring about, it also makes it slow to reverse once in train.

Global warming poses many difficult issues for international affairs, since it raises questions of ecological damage, differential economic effects among countries, equity between rich and poor nations, dependence on uncertain scientific calculations, and a host of other tangible and intangible issues. It will undoubtedly be an important source of both conflict and cooperation in the international system for many years into the future, and significant evolution will take place in international activities and institutions; but the issue is not likely to lead to substantial movement away from the existing nature of the international system as long as the effects are expected to be gradual and not catastrophic.

If, however, the threat turned out to be, or were perceived to be, real and catastrophic, it would pose new and possibly overwhelming problems for the international system. In that case, the continued assumption of autonomy on the part of nation-states would necessarily be challenged by the requirement for a level of collective action that went beyond any the international community has yet had to contemplate. The effects on the structure of the international system would have to be substantial. As we saw in the earlier discussion, the danger level remains far from that point as yet, though there are some voices that disagree.

What the effect would be on the international system if the threat from global warming or from other global dangers were accepted as catastrophic in scale cannot now be foretold. The usual assumption is that the overriding requirement for collective action would result in a major shift in the locus of power to international institutions and a significant decline in the authority of national governments. But it is also possible, even likely, that the result of such a challenge to the international structure would be a condominium of major powers exercising traditional hegemonic power over other states, rather than new forms of cooperative world government.

Thus, we see an international political system that has evolved substantially as a result of dynamic interaction with the outcomes of science and technology. Many elements of international affairs have been materially altered—autonomy, dependency, trade, economic competition, and the nature of warfare being perhaps the most immediately visible. The extent of evolution is substantial, but the general conclusion is as we have noted before: only a very few elements can be said to have changed in fundamental respects. Those few are important; they are concerned with the effect of nuclear weapons on the traditional role of military force and with the effects of new technologies, particularly information technologies, on the viability over time of centralized political power and on the competitiveness of centralized economic structures. As significant as these are, they have not yet altered, nor are they likely to alter, the underlying organizing principles of the international political system.

We have also seen, however, that some new issues and relationships tied to science and technology may come to challenge the existing system. The most significant candidate is the threat of substantial global warming as a result of human-produced effluents released into the atmosphere. If the threat proves to be catastrophic in scale, it could lead to a fundamental reordering of interstate relationships, or it could, perhaps more likely, lead to a reassertion of traditional patterns of state behavior.

The Processes, Outcomes, and Consequences of Science and Technology

In chapter 2, we explored the processes inherent in the operation of the scientific and technological enterprises, and discussed a number of natural outcomes to be expected from them. We have seen throughout the analysis how these processes and outcomes prefigure the character of the interaction with international affairs, whether it be the inevitable spread of information, the flexibility accorded to policy by the expansion of choice and opportunities, or other natural products of scientific and technological development. Of course, none of those processes or outcomes have effects independent of other factors; but they are significant inputs to international affairs, and will continue to be as long as the scientific and technological enterprises around the world remain more or less similar to what they are today.

We also asked whether these natural outcomes, as they interact with other societal factors, produce consequences for the organization and functioning of society that may be to some degree predictable—that is, whether there is a bias or trend that tends over time to favor certain consequences rather than their alternatives, even though the alternatives may, in the abstract, appear equally plausible. On the basis of this study, only a small number of such favored consequences could be confidently asserted, but those are important.

Open or Closed Societies

One of the most interesting questions is whether new technologies in general, and information technologies in particular, support open societies (those with essentially unfettered movement of information and ideas) over closed societies (those with rigid prohibitions and control exercised by the regime in power). The analysis in chapter 4 draws the conclusion that new technologies, on balance, *will* lead to a loosening of the bonds of a tightly controlled society rather than contribute to the continued viability of the controls. Though technology can certainly be used to constrain the movement of information and ideas, especially technology developed expressly for that purpose, that effort will be dominated eventually by the pressures for openness that new technology generally fosters as it is introduced and diffused. This tendency is an aspect of technology-induced change that is quite new; it is primarily a product of the stunning developments in information technologies in recent decades that contribute to their ubiquity, low cost, ease of use, and psychological impact.

Concentrated or Decentralized Political Power

Closely related is the question of concentration or decentralization of political power. Our conclusion, though more tentative than the first, is that the same effects of technology that contribute to the undermining of a closed society also contribute to the undermining of long-lasting authoritarian concentrations of political power. To put it positively, technological change will tend, on balance, to favor decentralization of political power within societies over time. Concentration of political power remains possible, but technological change tends to make it harder to preserve that concentration.

Centralized or Decentralized Economic Structures

We have also seen that a decentralized economic structure is far more congenial than a centralized one to the processes required for an economy to be technologically competitive. As a result, command economies cannot compete over time with decentralized market economies in fields requiring innovative capacity in science and technology.

Concentrated or Diffused Military Power

The consequences of technological change for concentration of military power are more ambiguous. The spread of weapons, and of the knowledge to produce them, is inevitable, thus implying unavoidable diffusion of power. But intensive concentrations of military power in a number of states can also continue. Thus, the general tendency toward diffusion of military power does not preclude continued or even enhanced concentrations of power in states rich enough and determined enough to make the effort.

Our analysis has not shown any other consequences of technological change that demonstrate consistent biases. In that sense, technology is neutral as far as other generalizable consequences are concerned; actual effects are determined instead by the interactions with economic, political, and social factors, with no specific trends attributable to technology observed.

Theoretical Debates

We have painted a picture in this work of an international system that is in considerable flux under the influence of technological change, with substantial evolution in the elements of international relationships, but in which only a few of the underlying concepts of international politics appear to have changed in definitive or fundamental ways. That picture may fly in the face of what seems to be commonly observed; certainly it does not agree with much of the rhetoric devoted to acclaiming the significance of the rapidity and extent of change.

Scholars in various theoretical schools of international relations are less inclined than popular commentators to easy acceptance of notions of fundamental change.[6] Science and technology, though recognized as the source of much of the evolution in international affairs in this century (particularly in the postwar era), are rarely examined in any detail in order to understand just how those fields interact with international affairs and how they are themselves influenced by the workings of international politics. As a result, science and technology are implicitly treated as static rather than dynamic variables; there is thus a tendency to miss, or to examine only piecemeal, how technological change affects (or does not affect) the constituent elements of different theoretical approaches. The role of science and technology may or may not be central to a judgment of the extent of change in international affairs that has already taken place or that may be in store; but that judgment cannot be made without probing science and technology more deeply than is typical in international-relations scholarship.

By and large, we did not try in the body of this work to relate the ideas being studied to overarching theoretical paradigms. Our goal was an examination—with careful attention to assumptions and testing of propositions—of this commonly missing element in international-relations theory, an examination that would remain independent of those broader paradigms and yet prove useful to their continued development.

Of course, it is not possible to be completely independent of preconceived notions or ideas. Our approach has a strong affinity with the perspective associated with the liberal theoretical school that perceives a complex web of interdependent and independent relationships existing along with, and to a considerable degree independent of, state-to-state relations. The very idea that the scientific and technological enterprises produce outcomes not necessarily sought by states, and that those outcomes will affect the power and influence of states, implicitly defines a world in which states are not in full control, in which there are independent elements at work influencing the play and evolution of world politics.

The realist view, on the other hand, which at first appears to be at odds with the influence the scientific and technological enterprises have on the evolution of the international system, may not be so easily dismissed on the basis of the evidence of our analysis. The realist perspective sees states as unitary actors in an anarchic international system, with power as the fundamental ordering principle among them—"a world of sovereign states seeking to maximize their influence and power."[7] A modified form of the realist perspective—sometimes called a neorealist or modified structural position—allows for a more disaggregated perspective on the role and influence of power, and recognizes the existence of international regimes outside governments as a necessary way for fully sovereign states to achieve desired outcomes in an interdependent world.[8]

What we have seen as a result of exploring the various influences of scientific and technological change is a mosaic in which the idealized constructs of a realist world are becoming increasingly anachronistic, but in which countervailing tendencies that are also influenced by science and technology could at the same time continue to support those constructs.

For example, it is clear that states are experiencing increasing constraints on their autonomy and authority as advances in technology encourage the growth of nonstate organizations and activities and the global integration that have become such prominent characteristics of international affairs. But science and technology also offer an increasingly large menu of options by which states can retain their power, and even enhance it, relative to competing institutions. Similarly, military power, long assessed as the key determinant of a state's power and influence, can no longer protect a state in a war-fighting role in confrontations between nuclear-armed states. The previously dominant role of military power in relationships among industrialized countries has thus been greatly altered, in favor of other sources of power and influence. Military power, however, is still relevant as a deterrent, and continues in its traditional role in international affairs (outside of nuclear confrontations)—supported by the continued progression and diffusion of conventional-weapons technology throughout the world. Other examples could be drawn from our analysis, but there are none that would disprove the proposition that states remain the dominant structural element in the international system.

It may be that only as we look to the future do we see developments related to science and technology that might more fundamentally challenge the realist conception by requiring a level of cooperation inconsistent with independent, sovereign states. The possibility of genuine threats to the entire planet's environment, the vulnerability of large systems on which the international economy depends, or the spread of weapons technology to an extent that poses intolerable threats to all countries, are developments that could precipitate a quite-different approach to the organization and functioning of the international system. Even then, the response to these dangers might be hegemonic rule by one or more states rather than willing cooperation. In fact, it would be quite rational to argue that the ability of the nation-state to mobilize political power, assemble resources, and organize effective action could well be the only realistic hope of meeting such threats.

Thus, it would be hard to assert on the basis of this evidence that the effects of scientific and technological change overwhelmingly undercut a pure or modified realist view of the world. Nor are the effects inconsistent with the liberal view of the international system. Sovereignty remains the organizing principle, though its content may be considerably eroded. Nation-states continue to be the dominant elements of the system and continue to strive to maximize their power to serve national purposes, though their authority and autonomy are becoming more constrained. Economic issues increasingly dominate international relationships, reducing, though not entirely replacing, military-power issues as the coin of the realm.

Neorealists might argue that the "reduction" of sovereignty is not meaningful, by definition, in a world of nation-states; or that military force has not been and cannot be minimized as the *sine qua non* that ultimately determines relations among states. Liberals might in turn want to argue that sovereignty is no longer the organizing principle of the international system, given the extent and growing importance of the nonstate organizations and activities that exist in parallel with nation-states.

The point for us is that the international system as we see it evolving under the influence of technological change can be accommodated within the constructs of both of those traditions. Many detailed aspects of the system have changed and are still changing; some of its long-standing concepts have been substantially altered. But the underlying structure of the international system appears basically similar to what it has been, whatever theoretical paradigm is used to describe it.

James Rosenau vehemently argues the contrary:

> It has proven to be much easier either to sweep the problem of new theory under the rug of "interdependence"—a word that implies that everything is connected to everything else, thus obscuring the need to discern underlying structures and processes—or to assert that nothing has changed because the anarchical arrangements which presently mark the conduct of world politics are much like those that prevailed during earlier eras. Yes, the power of states has eroded and yes, the world is smaller and more interdependent than ever, but the state is still the predominant actor in world politics and the state system is still the main foundation for the course of events. That has been the litany in the literature of the field for several decades."[9]

Rosenau's tone expresses the frustration he feels with current theoretical debates; he goes on to define in detail what elements he believes have evolved so substantially as to constitute a systemic change. He sees four new patterns: 1) the emergence of two overlapping worlds of global politics, consisting of a multi-centric world of diverse, relatively equal actors that is parallel to, and largely independent of, the state-centric world; 2) the diminished utility of force in the politics of the multi-centric world; 3) the elevation of autonomy as the driving force in the multi-centric world while security remains dominant in the other, although acquiring and preserving a "proper share of the world market" has come to rival acquisition of territory as a motivating force in the behavior of states; and, 4) most significantly in his analysis, the increased importance of individuals' skills, which are leading to the replacement of traditional criteria of legitimacy and authority for assessing the relative position of states with criteria of performance based on the skills of the populace.

The picture of the world that emerges in the pages of this study is not inconsistent with the one Rosenau paints, with similar observations about the growth of activities outside governments, changes in the utility of some levels of force, the growing shift to economic instead of military determinants of state power and influence, and the importance of technological skills. Whether to describe these as changes in kind or only changes of degree may then be largely a matter of definition, or of temperament.

However, where there are differences in the analysis that contribute to the differing assessments of systemic change, they are in the role of science and technology. Rosenau sees that role as a major causal factor of change. But he views it as a given, as exogenous to the international system, presenting it without appreciable critical examination and with the presumption that the implications of continued scientific and technological change are fairly obvious and consistent.

From our analysis, we have seen that science and technology are not exogenous but are to a degree sensitive to social choice; we have also seen that their implications are not necessarily obvious and will not always lead in the same direction. Three lessons stand out. First, technological change does not always have clear or consistent societal consequences; as summarized earlier in this chapter, in some cases there are dominating tendencies (e.g., change that tends to undermine closed societies), but in others the implications are less certain or may be reversible (e.g., the possibility of both centralization and dispersion of military power). Second, even when there are dominant tendencies in the consequences of technological change, determined government policies can have a substantial influence, especially in the short or medium term. Finally, technological change itself is not the independent variable it is often assumed to be; that is, governments can focus resources to attempt to produce desired technological outcomes. In fact, the motivations of foreign policy account for a substantial portion of the governmental resources devoted to targeted r/d.

It cannot be said that the world Rosenau describes is inaccurate, only that the dynamics of change are not so convincingly in one direction only. Thus, overall assessments of the extent of change are necessarily indeterminate for a system propelled in often-unpredictable directions by science and technology.

Moreover, it would not be difficult to construct a scenario in which the emergence of major challenges to the planet or to a large part of human society led to much greater centralization of authority in the hands of a few states in the international system. That authority might not be based on military power, but it could be based on a combination of military and economic power that would, in effect, constitute a reversal of today's trend toward a world of decentralized power.

Thus, we find that the military, economic, and social changes clearly stimulated or made possible by science and technology neither undercut nor strongly support the major theoretical paradigms current in international-relations theory though, on balance, we would have to say that they are much more consistent with liberal than with realist perspectives. But as we look ahead to the kinds of challenges the international system will face in the future, challenges in which technological change will play a major role, we could see either paradigm being fulfilled: greater cooperation among states to meet the challenges, with a gradual withering of the significance of state power in comparison to the power of nonstate institutions; or a reaffirmation of an anarchic system, dominated by a few powerful states. What is much less ambiguous is that theorists must more adequately recognize the dynamic and subtle nature of the interaction of science and technology with international affairs.

Comments on Policy

The issues discussed throughout this study have many implications for national and international policy, but the primary purpose has been descriptive and analytical—to assess the elements of change. Policy preferences were expressed with regard to specific subjects, but only in passing. Some implications are more or less self-evident, but most policy choices that arise in the context of the technology-induced evolution of international affairs are complex and controversial and cannot be briefly analyzed in proper detail. It would take a second book of at least equal length to discuss them in any depth; that would go far beyond our primary objective.

If our overall assessment had been different, there might have been more-dramatic overriding injunctions for policy initiatives. However, the conclusion that there has been substantial evolution in international affairs but little fundamental change in underlying concepts, and limited need for such change, does not suggest startling or novel policy prescriptions. Looking ahead, we have indicated several quite-diverse possibilities for the international system under the continued influence of developments based on science and technology. A revolution in the structure of international politics is certainly possible, but so, too, is a continuation of roughly the same pattern.

It would therefore be inappropriate to condition policy today as though a fundamental reorientation of the international system were certain or needed. Even if it occurs, it is at least several decades away and will in any case take a form that is at present quite unknowable.

We could ask whether evolution could be significantly influenced in a "desirable" direction by means of aggressive and ambitious policies that led toward some alternative international structure. Many take the position that the world is in sufficient trouble, whether it be poverty or hunger or ecological damage—even without the added possibility of planetary-level threats—that it is mandatory for enlightened governments to take an aggressive policy stance to alter the structure before it is too late. Usually, this position takes the form of a call for more transfer of authority to international institutions at the expense of nation-states, and for states to show a greater willingness to engage in collective action in which those institutions play a key role.

There can be no doubt, in fact, that the strengthening of international institutions and the restructuring of the policies of many industrialized nations toward these institutions must be accorded high priority, given the many requirements for their increased operational roles today and in the future. Whether such a general policy move should or would lead to a genuine restructuring of the international system, however, is quite another matter. It is not likely that the delegation of authority to international institutions would proceed very far once important interests of major states were adversely affected, nor would such a policy be seen as "desirable" by those who disagreed with the values and decision processes that might dominate the institutions.

Perhaps sometime in the future that degree of evolution will take place, but, to judge from the analysis of this study, the need for it has not yet been demonstrated; and the preferred form of system transformation, if there is a preferred form, is by no means obvious.

That is not meant as an argument for inactivity on meeting important international needs, or as a justification for lack of leadership. Much needs to be done, not least the strengthening of policies and institutions for effective collective action. Some of the other needs are no less familiar; some of the more important are easy to highlight. It is imperative that there be more concerted action to limit the diffusion of weapons and continued efforts to control and reduce armaments of mass destruction and their associated systems. States must be better prepared than most are at present for the scale and intensity of international cooperation that will be required on technology-intensive issues. Better mechanisms for encouraging third-world development and for the transfer of technology and resources from North to South are badly needed. Population growth, a serious negative factor in economic development and a contributor to wider international problems, must not continue unchecked; better population policies are required, along with more commitment to development and to improvement of public health. And critical attention must be given to emerging global threats and to the international functions that will be needed to meet them.

To enable action on these and other pressing matters, publics and governments must be better able to deal with the growing number of issues that involve science and technology directly and centrally. That requires, *inter alia*, more attention to improving the skills of the population at large so that it can understand and respond to the increasing scientific and technological content of daily life.

These and other policy needs, to some degree all related to science and technology, form a challenging list when moved from the banality of general statements to the detailed policy analyses necessary to make something happen. All appear on the national and international agendas in one form or another. That, however, does not mean that they are receiving adequate attention, nor does it diminish their increasing importance.

Coda

It is a comfortable, perhaps a complacent, conclusion that scientific and technological advance has led to substantial and continuing evolution in international affairs but has not led to more than limited challenges to the traditional assumptions and concepts that guide nations in their international relationships. Drastically new modes of thinking about international affairs are not required, nor need we worry about the consequences of an imminent restructuring of the international political system. As just noted, that conclusion is no guarantee that the policies that *are* sorely needed will be forthcoming; but as serious as the deficiency in adequate policies may be, it is unfortunately a familiar problem of governance.

I must admit to some disquiet, however, on several scores. One is simply the realization that the magnitude of human needs and the sheer difficulty of effective governance in a complex world make optimism about the human condition difficult to sustain. Another is a disturbing sense that we have come to rely excessively on science and technology not only to understand the world, but also to provide the path for rescue from whatever trouble in which we find ourselves.[10] As should be clear from the many examples in this book, science and technology can be used to contribute to the amelioration of many serious social problems and sometimes to find a way of bypassing them. But whether the problems can be effectively dealt with will depend on the interaction of many factors, never on science and technology alone.

And beyond that is the disquiet that arises from consideration of possible future products of the scientific and technological enterprises and of their societal effects. At the beginning of this chapter, the observation was made that the dramatic developments in science and technology during the past century (even the past several centuries) and their consequences for society have taken place over an astonishingly small portion of humanity's total experience. Technologies we now take for granted were in many cases not even dreamed of only a very few decades or even years ago; the same can be said for some of the widespread consequences of technological change. There is no reason to believe that that pace will slacken, a fact that should lead to some skepticism toward any perspective that pictures the future as roughly similar to the present.[11]

Advances already under way in many fields, and countless others not yet even thought of, will inevitably lead to major changes in available technologies. Many of these will have direct effects on international affairs through the patterns of outcomes we outlined in chapter 2—for example, by multiplying the options for national policy or by contributing to the diffusion of power; all are likely to have an impact, perhaps a much larger one, through the societal changes that will follow their introduction. The nature of human society, even some four or five decades from now—a very brief time in terms of human history—could conceivably be strikingly different from that of the last decade of the twentieth century, with the international system as affected by the differences as any other element.

As we look at current progress in science and technology, it is impossible not to be awed by advances in a wide range of fields. Molecular biology is a prime illustration. The cascading expansion of knowledge about the genetic code and about biological processes promises powerful new technologies for intervention in disease, hereditary defects, and agriculture; for production of new materials; and for new production processes. Similarly, new technologies for miniaturization of materials and devices, just taking shape but already allowing the manipulation of matter atom by atom, are opening an entire new physical realm, with possible applications limited only by the power of the imagination.[12] Computers continue to become faster, smaller, and less expensive, and their capacity and possible applications increase correspondingly. Research and development on materials increasingly makes possible design on demand, which allows both development of technologies able to operate under previously impossible conditions and substitution for costly or scarce resources.

The possible illustrations are manifold; but it would be fruitless, and quite misleading, to attempt to highlight all the scientific and technological advances of today that will have an impact over several decades. They are not all imaginable, nor could those identified be calibrated as to their importance. Moreover, the most significant aspects for the international scene will be not the technologies themselves, but the societal effects as they unfold over time.

In that respect, as we try to formulate ideas about the world of just a few decades off, it is striking, and disturbing, that society is still trying to cope with the profound effects of the scientific revolutions that, beginning in the sixteenth century, first removed human beings from their central role in the cosmos, then reduced them to just one more cog in an impersonal evolutionary scheme, and finally gave them, in this century, the power actually to destroy the heritage of millennia, even to destroy the human race entirely. Most recently, humans have been able to view their habitat as the isolated, apparently fragile ball in the universe it actually is; the astronomer Fred Hoyle said in 1948, "Once a photograph of the Earth, taken from outside, is available, once the sheer isolation of the Earth becomes plain, a new idea as powerful as any in history will be let loose."[13] How important these changes in the philosophical underpinning of society actually are is a continuing field of inquiry; but over time these changes must, in an important sense, be central to the conception of human beings' relation to the environment and to one another.

Now, two additional fundamental advances in knowledge are in the offing, advances that are likely to have equally profound effects on humanity's view of itself and of its place in the universe.

The first is the unraveling of the origin and structure of life, perhaps even including the creation of life from inert compounds. This knowledge will certainly offer major beneficial technologies, but it will present wrenching moral and policy dilemmas. It will also further challenge the psychological and philosophical foundations of traditional views of the inviolability and uniqueness of human life and of its genetic inheritance.[14]

The other is the detailed understanding of the functioning of the brain—knowledge that could make possible not only intervention in an individual's mental processes, but also replication of the brain in the design of computers and thus, possibly, machine recreation of something akin to human consciousness. It may also eventually make possible the introduction of knowledge through direct transmission of electrical signals to the brain. If any of these prospects came to fruition, what would be implied about the practical consequences of the possibilities of control of individuals, the generation and analysis of information, and the augmentation of the intellectual capacity of human beings is simply unknown and unknowable today. What is certain is that any such capabilities would be bound to have profound effects on fundamental conceptions of humanness and humanity—and quite likely would increase the prevalence of anti-science and anti-technology attitudes, already substantial in society at large.[15]

These two developments may not come about for decades, or they may never come about; but they are not at all beyond the realm of possibility. Their effects cannot be adequately anticipated, but they are certain to have an important psychological and social impact akin to and perhaps more tumultuous than, that of the earlier revolutionary scientific advances. The effects will not be only in the international or global sphere, but they certainly will be felt there. How they will influence the evolution of international politics and the structure of the international system is now far from being predictable. But the prospect of such developments makes it evident that our "comfortable" conclusion about the lack of fundamental change in the international system is a conclusion only for the present and for a few decades into the future; it could easily need to be reformulated as the wonders of the scientific and technological enterprises continue to produce their magic.

It is a commonplace that science and technology are both curse and savior. *The Economist* said it well:

> Mankind at the approach of the third millennium is more addicted to the drug of progress than ever, yet more eager to get rid of that addiction; more distrustful of those who would make better and better versions of that drug, yet more ready to pay them; more aware of the infinity of improvements in the drug that are still possible, yet more convinced that the drug is not the answer to its problems. Ignorance has shrunk because of science, but it has also grown: every discovery reveals how much is not known. [16]

Science and technology, the embodiment of the quest for knowledge and understanding, will continue to be central elements of society into the next century, with support for research coming from both private and public sources. Nations will have to cope with the results, trying to shape them for their purposes and to limit deleterious effects. Whatever form the evolution of international affairs takes in this larger framework, the one predictable constant will be change.

Notes _____

Chapter One

1. Within the now-large literature concerning the role of technology in world history, the works of William H. McNeill and Lynn White are among the most valuable, and most useful as references to other works. See particularly William H. McNeill, *The Pursuit of Power: Technology, Armed Force, and Society since A.D. 1000* (Chicago: University of Chicago Press, 1982); and Lynn White, Jr., *Medieval Technology and Social Change* (Oxford: Oxford University Press, 1962).

2. Don K. Price, *Government and Science* (New York: Oxford University Press, 1962); Louis Hartz, *The Liberal Tradition in America: An Interpretation of American Political Thought since the Revolution* (New York: Harcourt, Brace & World, 1955).

3. George P. Shultz, address before the World Affairs Council, Dec. 4, 1987, published in U.S. Department of State, Bureau of Public Affairs, *National Success and International Stability in a Time of Change*, Current Policy no., 1029, p. 1; W. Michael Blumenthal, "The World Economy and Technological Change," *Foreign Affairs* 66, no. 3 (*America and the World*, 1987/88): 531.

4. See particularly Lloyd Berkner, *Science and Foreign Relations: International Flow of Scientific Information*, International Science Policy Survey Group, U.S. Department of State (Washington, D.C., 1950); Eugene B. Skolnikoff, *Science, Technology, and American Foreign Policy* (Cambridge, Mass.: MIT Press, 1967); U.S. House of Representatives, Committee on International Relations, Subcommittee on International Security and Scientific Affairs, *Science, Technology, and Diplomacy in the Age of Interdependence*, June 1976; John V. Granger, *Technology and International Relations* (San Francisco: W. H. Freeman, 1979); Joseph S. Szyliowicz, ed., *Technology and International Affairs* (New York: Praeger, 1981); Ralph Sanders, *International Dynamics of Technology* (Westport, Conn.: Greenwood Press, 1983); and Kenneth H. Keller, "Science and Technology," in Nicholas X. Rizopoulos, ed., *Sea-Changes: American Foreign Policy in a World Transformed* (New York: Council on Foreign Relations, 1990).

5. James N. Rosenau offers a particularly insightful discussion of the concept of change and the difficulties of assessing change in global affairs in "Global Transformations," Occasional Paper No. 1, International Security Studies Program, American Academy of Arts and Sciences (Cambridge, Mass., Nov. 1989).

6. James N. Rosenau, *Turbulence in World Politics: A Theory of Change and Continuity* (Princeton: Princeton University Press, 1990). This book is a fascinating and lucid approach to defining, with both clarity and some precision, a means to assess the extent of change in what Rosenau calls postindustrial politics. His argument is necessarily still subjective and definitional, though he presents supporting data. It is, however, a more elaborated and closely reasoned statement of fundamental change in the system than any other of which I am aware; it is discussed further in chapter 7.

7. Ibid., p. 97. At various places Rosenau indicates his impatience with those who see continuity rather than change: "Some analysts derive comfort from feeling that the present is but a continuation of the past" (p. 70), and "It is like arguing that, despite an ever-expanding global interdependence, little has changed in world politics because states still conduct foreign policy" (p. 74).

8. Many current theoretical issues in international relations must deal with the implications of science and technology—for example, the development and signif-icance of international regimes, or of interdependence. But, typically, the analyses do not probe science and technology in any depth. Even Rosenau's *Turbulence in World Politics* tends to present technology as a given rather than as a dynamic variable in its own right.

9. A speech by former Secretary of State Shultz is typical: "We have a race between the engineer and the politician, the creators of new knowledge, and the statesman: for the very idea of a nation, the concept of national sovereignty, is affected. . . . as national boundaries blur, sovereign power is dispersed, and new players vie for international influence, the time-honored concept of sovereignty needs to be reexamined" ("On Sovereignty," speech given on the occasion of the twenty-fifth anniversary of the National Academy of Engineering, Washington, D.C., Oct. 4, 1989).

10. Stephen D. Krasner, "Regimes and the Limits of Realism," in Stephen D. Krasner, ed., *International Regimes* (Ithaca, N.Y.: Cornell University Press, 1983), p. 366.

11. Alan James, *Sovereign Statehood: The Basis of International Society* (Lon-don: Allen & Unwin, 1986). What others might refer to as erosion of sovereignty, James would refer to as growing constraints on the freedom of action of a sovereign state.

12. A precise and agreed definition would not, in fact, be fully achievable, as the many and varied attempts at definition attest. For some useful approaches, see Langdon Winner, *Autonomous Technology* (Cambridge, Mass.: MIT Press, 1977); Granger, *Technology and International Relations;* Szyliowicz, *Technology and In-ternational Affairs;* and Harvey Brooks, "Technology, Evolution, and Purpose," in "Modern Technology: Problem or Opportunity?" *Daedalus* 109, no. 1 (Winter 1980): 65–81.

13. Brooks, "Technology, Evolution, and Purpose," p. 66.

14. This does not imply that all technological knowledge need be able to be codified in formal descriptions or formulas. There is some knowledge, often referred to as "embedded" knowledge or "know-how," that is embodied in the experience of an individual or organization that has gone through the process of producing a piece of hardware. Such knowledge, too, is legitimately "specifiable and reproducible."

15. Brooks, "Technology, Evolution, and Purpose," p. 66.

16. Of course, innovation itself can also be considered a technology when and if it is reduced to a set of specified rules and procedures. Industries and nations are trying to do just that. But for now the overlap of definitions can be treated as a second-order complication not essential for the points being made.

17. Brooks, "Technology, Evolution, and Purpose," p. 66.

18. Economics is a partial exception, for its relationships to policy are often

similar to those of the hard sciences; and in practice it is more likely to be the basis of policy than are other social sciences. Though we will not probe in detail the history of economics as a discipline, the role of economics will emerge more often in the text than will those of other social-science disciplines.

Chapter Two

1. The literature on the development of science is rich and vast. A recent comprehensive work with extensive references to other literature is Daniel J. Boorstin, *The Discoverers* (New York: Random House, 1983). A succinct summary of the role of Newton and the Royal Society is presented in Harvey Brooks, "Can Science Survive in the Modern Age?" *Science* 174 (Oct. 1, 1971): 21.

2. There is also a vast literature on the history of technology. See in particular Lynn White, Jr., *Medieval Technology and Social Change* (Oxford: Oxford University Press, 1962); William H. McNeill, *The Pursuit of Power: Technology, Armed Force, and Society since A.D. 1000* (Chicago: University of Chicago Press, 1982); Arnold Pacey, *Technology in World Civilization: A Thousand-Year History* (Cambridge, Mass.: MIT Press, 1990); and Thomas Parke Hughes, *The Development of Western Technology since 1500* (New York: Macmillan, 1964).

3. A wealth of insight into the role of science and technology in the economic growth of the West is presented in Nathan Rosenberg and L. E. Birdzell, Jr., *How the West Grew Rich: The Economic Transformation of the Industrial World* (New York: Basic Books, 1986); it will be relied on heavily in this chapter. See also David C. Mowery and Nathan Rosenberg, *Technology and the Pursuit of Economic Growth* (New York: Cambridge University Press, 1989).

4. There is an extensive literature on the question of the forces that led to the advance of technology; a particularly useful summary view and references to other key works are found in Rosenberg and Birdzell, *How the West Grew Rich*.

5. Paul Kennedy, *The Rise and Fall of the Great Powers: Economic Change and Military Conflict from 1500 to 2000* (New York: Random House, 1987), p. 17.

6. Ibid., p. 439.

7. Rosenberg and Birdzell, *How the West Grew Rich*, p. 244.

8. McNeill, *Pursuit of Power;* Jacques S. Gansler, *Affording Defense* (Cambridge, Mass.: MIT Press, 1989); Elting E. Morison, *From Know-how to Nowhere: The Development of American Technology* (New York: Basic Books, 1974).

9. There are many accounts of the wartime application of science and technology. See especially James Phinney Baxter 3d, *Scientists against Time* (Boston: Little, Brown, 1946); and Robert L. O'Connell, *Of Arms and Men: A History of War, Weapons, and Aggression* (Oxford: Oxford University Press, 1989).

10. A. Hunter Dupree, *Science in the Federal Government: A History of Policies and Activities to 1940* (Cambridge, Mass.: Harvard University Press, Belknap Press, 1957), p. 26.

11. Ibid., pp. 149–83; Thomas M. Arndt, Dana G. Dalrymple, and Vernon W. Ruttan, *Research Allocation and Productivity in National and International Agricultural Research* (Minneapolis: University of Minnesota Press, 1977).

12. *OECD in Figures: Supplement to the OECD Observer* (Organization for Economic Cooperation and Development, Paris), no. 170 (June/July 1991): 52–55.

13. Precision in these figures is not possible, notwithstanding the extensive data bases in some countries, for the data is bedeviled by problems of definition, perspective, and simple absence of information. It is also possible that commitments as they were at the end of the 1980s could undergo substantial revision as a result of the dramatic collapse of the threat from Eastern Europe. The changes will take place only gradually, however, with economic constraints (recessions or worse) and other slow-moving shifts in social preferences likely also to have a substantial impact on r/d expenditures.

14. The figure for science and technology activities as a whole would be much larger than this, for it would include items not captured in r/d data such as production, application, and marketing of technology. Only r/d data is available, and in any case it is the most relevant to the development of new technology.

15. *OECD in Figures*, pp. 52–53.

16. Figures for the USSR not only are difficult to obtain but are subject to many uncertainties, including the appropriate exchange rate to use. Loren R. Graham has estimated expenditures on science (presumably basic and applied research) as averaging more than 26 billion rubles per year in the early 1980s ("Science and Technology Trends in the Soviet Union," Program in Science, Technology, and Society, Massachusetts Institute of Technology, Cambridge, Mass., Oct. 1, 1987). That would be approximately $40 billion at official rates for that time, implying total r/d expenditures of some $120 billion (assuming that basic and applied research accounts for about one-third of the total, as it does in government funding in the United States).

17. National Science Foundation, *Science and Technology Data Book—1990*, NSF 90-304 (Washington, D.C.: June 1990), p. 20.

18. Ibid., p. 32.

19. American Association for the Advancement of Science, Intersociety Working Group, *AAAS Report XVII: Research and Development, FY 1993* (Washington, D.C., 1990), p. 50.

20. Roughly, basic research seeks knowledge of fundamental processes, without necessary relevance to application; applied research seeks knowledge relevant to an application known in advance; and development is the application of knowledge for a specified purpose.

21. Harvey Brooks, "Technology, Evolution, and Purpose," in "Modern Technology: Problem or Opportunity?" *Daedalus* 109, no. 1 (Winter 1980): 68–70.

22. Ibid., p. 69.

23. Henry Ergas, "Does Technology Policy Matter?" in Bruce R. Guile and Harvey Brooks, eds., *Technology and Global Industry: Companies and Nations in the World Economy*, National Academy of Engineering (Washington, D.C.: National Academy Press, 1987), pp. 192–245. The material in the following paragraphs is drawn from this seminal work. A useful and succinct analytical description of U.S. structure and policies for science and technology is found in David C. Mowery and Nathan Rosenberg, "The U.S. National Innovation System," CCC Working Paper No. 90-3, Center for Research in Management, University of California at Berkeley, Sept. 1990.

24. Gansler, *Affording Defense*, pp. 141–214.

25. The pressures in the USSR were similar; but the Soviet Union, because of

its lag in technology as well as its different historical experience, placed more emphasis on developing conservative technology adequate for its purposes, rather than being totally committed to the state of the art. But it also believed it was necessary to match the pace of U.S. innovation in military-related technology in order to remain competitive in strategic capabilities. The effort to maintain technological parity proved to be an increasingly difficult task, and it ultimately contributed to the massive changes in policy and structure of the Gorbachev era and to the collapse of the Communist system. The underlying reasons for the inability of the system to maintain parity with the West are discussed in chapter 4.

26. The authors of a Harvard study argue that fast-paced incremental improvement in production and product characteristics constitutes one of the recent "revolutions" that have altered the nature of technological competition among nations (John M. Alic et al., eds., *Beyond Spinoff: Military and Commercial Technologies in a Changing World* [Boston: Harvard Business School Press, 1992], chap. 1).

27. David J. Teece, "Capturing Value from Technological Innovation: Integration, Strategic Partnering, and Licensing Decisions," in Guile and Brooks, *Technology and Global Industry*, pp. 65–95.

28. Daniel Clery, "The Greening of EUREKA," *New Scientist* 126 (June 9, 1990): 38; David Swinbanks, "Going for 'Green Technology,'" *Nature* 350 (Mar. 28, 1991): 266–67.

29. Giovanni Dosi gives a useful summary of theories of the determinants and direction of technological change, along with suggestions of his own, in "Technological Paradigms and Technological Trajectories," *Research Policy* 11 (1982): 147–62.

30. Nathan Rosenberg, *Perspectives on Technology* (Cambridge: Cambridge University Press, 1976), pp. 108–25.

31. Alic et al., *Beyond Spinoff*, chap. 4.

32. *Proceedings*, Symposium on World Telecommunications Policy: In Tribute to Ithiel de Sola Pool, Massachusetts Institute of Technology, Cambridge, Mass., Jan. 27–28, 1988.

33. Thomas P. Hughes, "Conservative and Radical Technologies," in Sven Lundstedt and E. William Colglazier, Jr., eds., *Managing Innovation: The Social Dimensions of Creativity, Invention, and Technology* (Oxford: Pergamon Press, 1982), pp. 31–44.

34. Alic et al., *Beyond Spinoff*, chap. 3.

35. Christopher Freeman, *Technology Policy and Economic Performance: Lessons from Japan*, Science Policy Research Unit, University of Sussex (New York: Pinter, 1987).

36. For a current model of innovation that relies heavily on the "chain-linked" model of repeated interactions between science and technology, see Stephen J. Kline and Nathan Rosenberg, "An Overview of Innovation," in Ralph Landau and Nathan Rosenberg, eds., *The Positive Sum Strategy: Harnessing Technology for Economic Growth* (Washington, D.C.: National Academy Press, 1986), pp. 275–305.

37. National Academy of Sciences, *Scientific Communication and National Security: A Report by the Panel on Scientific Communication and National Security*

(the Corson Report) (Washington, D.C.: National Academy Press, 1982); National Academy of Sciences, *Balancing the National Interest: U.S. National Security Export Controls and Global Economic Competition* (the Allen Report) (Washington, D.C.: National Academy Press, 1987). The issue of control of scientific and technological information and hardware is discussed in chapter 3.

38. "The International Relationships of MIT in a Technologically Competitive World," report of a faculty study group, Massachusetts Institute of Technology, Cambridge, Mass., May 1, 1991.

39. Alic et al., *Beyond Spinoff*, chap. 2.

40. The progress Iraq had made by 1991 toward developing nuclear weapons is an unfortunate illustration (William J. Broad, "Iraqi Atom Effort Exposes Weakness in World Controls," *New York Times*, July 15, 1991, p. 1; R. Jeffrey Smith, "Grim Lessons for the World in Iraq's Secret Nuclear Program," *Washington Post National Weekly Edition*, Aug. 19–25, 1991, p. 22; Michael Wines, "U.S. Is Building Up a Picture of Vast Iraqi Atom Program," *New York Times*, Sept. 27, 1991, p. A8).

41. David P. Hamilton, "Superconductor Race Shifts to a New Arena," *Science* 250 (Oct. 19, 1990): 374–75; Office of Technology Assessment, *High-Temperature Superconductivity in Perspective* (Washington, D.C.: OTA, May 1990).

42. Henry de Wolf Smyth, *Atomic Energy for Military Purposes: The Official Report on the Development of the Atomic Bomb under the Auspices of the United States Government, 1940–1945* (Princeton: Princeton University Press, 1945).

43. The idea of accelerating change arises often in popular and scholarly writing, though usually without extensive analysis. See Henry Adams, *The Education of Henry Adams* (New York: Random House, 1931), chap. 34, "A Law of Acceleration"; and Arthur M. Schlesinger, Jr., "The Challenge of Change," *New York Times Magazine*, July 27, 1986, pp. 20–21.

44. M. D. Fagan, ed., *A History of Engineering and Science in the Bell System: The Early Years, 1875–1925* (Bell Telephone Laboratories, Inc., 1975), pp. 1–25; White, *Medieval Technology and Social Change*.

45. Paul Johnson argues that the most substantial changes from the past occurred during 1815–30, and that those years constituted the dawn of the modern era (*The Birth of the Modern: World Society, 1815–1830* [New York: Harper Collins, 1991]).

46. *Technology, Trade, and the U.S. Economy*, report of a workshop held at Woods Hole, Mass., Aug. 22–31, 1976 (Washington, D.C.: National Academy of Sciences, 1978), p. 23.

47. In fact, in some sectors, the time needed for full exploitation of a new technology may be longer than in the past. In the transportation sector, the time constant for substitution of new transport infrastructures has grown to 30 years for canals, 50 years for railways, 90 years for surfaced roads, and 130 years for aircraft (Nebojsa Nakicenovic, "Dynamics and Replacement of U.S. Transport Infrastructures," in Jesse H. Ausubel and Robert Herman, eds., *Cities and Their Vital Systems: Infrastructure Past, Present, and Future* [Washington, D.C.: National Academy Press, 1988], pp. 175–221).

48. Dorothy Ross, *The Origins of American Social Science* (New York: Cambridge University Press, 1990); Charles M. Bonjean, Louis Schneider, and Robert

L. Lineberry, *Social Science in America: The First Two Hundred Years* (Austin: University of Texas Press, 1976).

49. Gene Lyons, ed., "Social Science and the Federal Government," *Annals* (American Academy of Political and Social Science) 394 (Mar. 1971); Austin Ranney, ed., *Political Science and Public Policy* (Chicago: Markham, 1968).

Chapter Three

1. Hans J. Morgenthau, *Politics among Nations: The Struggle for Power and Peace*, 5th ed. (New York: Alfred A. Knopf, 1973); Kenneth N. Waltz, *Theory of International Politics* (Reading, Mass.: Addison-Wesley, 1979).

2. Lynn White, Jr., *Medieval Technology and Social Change* (Oxford: Oxford University Press, 1962); Bernard and Fawn Brodie, *From Crossbow to H-Bomb* (Bloomington: Indiana University Press, 1973); William H. McNeill, *The Pursuit of Power: Technology, Armed Force, and Society since A.D. 1000* (Chicago: University of Chicago Press, 1982).

3. Bernard Brodie, "Technological Changes, Strategic Doctrine, and Political Outcomes," in Klaus Knorr, ed., *Historical Dimensions of National Security Problems* (Lawrence: University Press of Kansas, 1976), p. 286.

4. James Phinney Baxter 3d, *Scientists against Time* (Boston: Little, Brown, 1946); Alan S. Milward, *War, Economy, and Society, 1939–1945* (Berkeley: University of California Press, 1977).

5. Robert W. Tucker, "1989 and All That," in Nicholas X. Rizopoulos, ed., *Sea-Changes: American Foreign Policy in a World Transformed* (New York: Council on Foreign Relations, 1990), p. 214. A particularly good review of the history, meaning, and problems of the nuclear era is to be found in Robert Jervis, *The Meaning of the Nuclear Revolution: Statecraft and the Prospect of Armageddon* (Ithaca, N.Y.: Cornell University Press, 1989).

6. See, for example, John Keegan, *The Face of Battle* (New York: Viking Press, 1976); John U. Nef, *War and Human Progress: An Essay on the Rise of Industrial Civilization* (Cambridge, Mass.: Harvard University Press, 1950); Carlo Cipolla, *Guns, Sails, and Empires: Technological Innovation and the Early Phases of European Expansion, 1400–1700* (New York: Pantheon Books, 1965); Martin L. Van Crevald, *Technology and War: From 2000 B.C. to the Present* (New York: Free Press, 1989); and McNeill, *Pursuit of Power*.

7. There are many variations on the rationale for building large nuclear forces, varying from arguments like those of Herman Kahn, who believed a nuclear war could be fought and won and that forces should be designed with that in mind (*On Thermonuclear War*, 2d ed. [Princeton: Princeton University Press, 1961]) to those of Bruce Russett, who believes that nuclear weapons have been used not to prevent war between the superpowers, but for "extended deterrence"—that is, to protect security interests against the other superpower, beyond the superpower borders ("The End of Nuclear Efficacy," *Arms Control Today* 17, no. 7 [Sept. 1987]: 37–38). Important as these debates are, this is not the place to explore their ramifications, nor is it necessary for our purposes.

8. "Estimated Strategic Forces under START," *Arms Control Today* 21, no. 3 (Apr. 1991): 30.

9. Some of the many influential works on nuclear weapons and deterrence are John H. Herz, *International Politics in the Atomic Age* (New York: Columbia University Press, 1959); Albert Wohlstetter, "The Balance of Terror," *Foreign Affairs* 37, no. 2 (Jan. 1959): 211–34; Herman Kahn, *Thinking about the Unthinkable* (New York: Horizon Press, 1962); Henry Kissinger, *Nuclear Weapons and Foreign Policy* (New York: Harper & Brothers, 1957); Ashton B. Carter, John D. Steinbruner, and Charles A. Zraket, eds., *Managing Nuclear Operations* (Washington, D.C.: Brookings Institution, 1987); and Jervis, *Nuclear Revolution*.

10. "A nuclear war cannot be won and must never be fought" (Joint statement of President Ronald Reagan and First Secretary Mikhail Gorbachev, Nov. 21, 1985, as quoted in Jervis, *Nuclear Revolution*, p. 1).

11. Office of Technology Assessment, *Strategic Defenses, Ballistic Missile Defense Technologies, Anti-Satellite Weapons, Countermeasures, and Arms Control* (Princeton: Princeton University Press, 1986), pp. 37–38. The Strategic Defense Initiative is discussed later in this chapter.

12. Joseph S. Nye, Jr., *Bound to Lead: The Changing Nature of American Power* (New York: Basic Books, 1990), pp. 69–70.

13. Paul Kennedy makes the general argument in his book *The Rise and Fall of the Great Powers: Economic Change and Military Conflict from 1500 to 2000* (New York: Random House, 1987). Joseph Nye responds with a different view in *Bound to Lead*, as does Henry R. Nau in *The Myth of America's Decline: Leading the World Economy into the 1990s* (New York: Oxford University Press, 1990).

14. Carl Kaysen, Robert S. McNamara, and George W. Rathjens, "Nuclear Weapons after the Cold War," *Foreign Affairs* 70, no. 4 (Fall 1991): 95–110; Barton Gellman, "Button, Button, Who's Got the Button?" *Washington Post National Weekly Edition*, Sept. 2–8, 1991, p. 11.

15. "A New Era of Reciprocal Arms Reductions," *Arms Control Today* 21, no. 8 (Oct. 1991): 3–6. A major arms-control agreement (START) between the United States and the USSR requiring reductions by approximately one-third in strategic weapons was signed in Moscow on July 17, 1991 (Eric Schmitt, "Senate Approval and Sharp Debate Seen," *New York Times*, July 19, 1991, p. A7).

16. John J. Mearsheimer argues the extreme position that the passing of the cold war will result in the return of old antagonisms in Europe, with the emergence of a multipolar state system—including nuclear arms—in which deterrence will be difficult to maintain ("Back to the Future: Instability in Europe after the Cold War," *International Security* 15, no. 1 [Summer 1990]: 5–56). His argument derives from pure realist theory and is hard to relate to actual evidence. See responses by Stanley Hoffmann and Robert O. Keohane, and a reprise by Mearsheimer, in *International Security* 15, no. 2 (Fall 1990): 191–99.

17. John A. Alic et al., eds., *Beyond Spinoff: Military and Commercial Technologies in a Changing World* (Boston: Harvard Business School Press, 1992), chap. 3.

18. Of course, many nuclear analysts and planners would take issue with this assertion; they would argue that it is a very much more complex issue, dependent on the details of force planning, on the scenarios of possible incidents, on the nature of a particular technological development and how quickly either side could exploit it, and on other detailed questions. But in my view, given anything approx-

imating force levels and configurations of the 1980s, or after the reductions mandated by the START agreement, adventurism by any nuclear power against another is entirely deterred, and there are no incremental changes that could alter that situation.

19. See the extended discussion of the symbolic aspects of nuclear politics in Jervis, *Nuclear Revolution*, pp. 174–225; and Philip A. G. Sabin, *Shadow or Substance? Perception and Symbolism in Nuclear Force Planning*, Adelphi Papers No. 222 (International Institute of Strategic Studies), Summer 1987.

20. This shortcoming in the former Soviet Union was particularly important in the slow pace of computer development. See S. E. Goodman, "Technology Transfer and the Development of the Soviet Computer Industry," in Bruce Parrott, ed., *Trade, Technology, and Soviet-American Relations* (Bloomington: Indiana University Press, 1985), pp. 117–40.

21. The classic case was the need for the Soviets to build rockets larger than those of the United States for their initial ICBM, since their nuclear warheads were more primitive and thus much heavier in relation to yield than U.S. warheads (Charles Maier, Introduction to George B. Kistiakowsky, *A Scientist at the White House: The Private Diary of President Eisenhower's Special Assistant for Science and Technology* [Cambridge, Mass.: Harvard University Press, 1976], p. xxxiii). As an unexpected dividend, the Soviet Union had a large advantage in lift capacity in early space launches and reaped political kudos because it appeared as a result to be technologically more advanced than the United States.

22. Other arguments were also used to support export controls, such as reducing U.S. r/d costs if the rate of change of Soviet technology was slowed, forcing higher r/d expenditures on the Soviet Union when it could ill afford them, and publicly emphasizing the technological gap between the U.S. and the U.S.S.R.

23. Mitchel B. Wallerstein and Stephen B. Gould, eds., "A Delicate Balance: Scientific Communication vs. National Security," *Issues in Science and Technology* 4, no. 1 (Fall 1987): 42–55. The extent of the Soviet Union's efforts to acquire Western technology is detailed in an anonymous report, presumably by the Central Intelligence Agency or the Defense Intelligence Agency, which was updated periodically (*Soviet Acquisition of Militarily Significant Western Technology: An Update*, Sept. 1985).

24. Peter Coles, "Easier Access," *Nature* 345 (June 14, 1990): 567. The likelihood of the spread of the latest technologies to additional nations has led to new concerns on the part of the U.S. Department of Defense—for example, about the use of computers for weapons design or for submarine detection (John Markoff, "New Curbs on Exports Are Sought," *New York Times*, Sept. 11, 1991, p. D1).

25. For the best analysis of this issue, see the two reports of the National Academy of Sciences and its sister institutions: *Scientific Communication and National Security: A Report by the Panel on Scientific Communication and National Security* (the Corson Report) (Washington, D.C.: National Academy Press, 1982); and *Balancing the National Interest: U.S. National Security Export Controls and Global Economic Competition* (the Allen Report) (Washington, D.C.: National Academy Press, 1987).

26. *Balancing the National Interest*, pp. 54–69, 116–26.

27. "Hard Pounding," *The Economist*, July 11, 1987, p. 64; Marshall Goldman, "The Case of the Not-So-Simple Machine Tools," *Technology Review* 90, no. 7 (Oct. 1987): 20; *Balancing the National Interest*, pp. 146–48.

28. These arguments were discussed in depth in *Scientific Communication and National Security;* and *Balancing the National Interest.*

29. Jeffrey T. Richelson, *America's Secret Eyes in Space: The U.S. Keyhole Spy Satellite Program* (New York: Harper & Row, 1970): B. Jasani, ed., *Outer Space: A New Dimension of the Arms Race* (London: Taylor & Francis, 1982); John C. Toomay, "Warning and Assessment Sensors," in Carter, Steinbruner, and Zraket, *Managing Nuclear Operations*, pp. 282–321.

30. Two serious incidents occurred in the U.S. systems in 1979 and 1980, one caused by human error in the playing of a test tape, the other by a faulty computer chip. Five false warnings have been cited in Soviet systems since 1982, all caused by incorrect interpretations of satellite intercepts, missile tests, and the aurora borealis (Scott D. Sagan, "Reducing the Risks: A New Agenda for Military-to-Military Talks," *Arms Control Today* 21, no. 6 [July/Aug. 1991]: 17).

31. William J. Broad, "Serious Sharing of 'Star Wars'? Not in This Millennium," *New York Times*, Feb. 23, 1992, p. E5.

32. For excellent discussions of command and control issues, see Bruce G. Blair, *Strategic Command and Control: Redefining the Nuclear Threat* (Washington, D.C.: Brookings Institution, 1985); Paul J. Bracken, *The Command and Control of Nuclear Forces* (New Haven, Conn.: Yale University Press, 1983); and Bruce G. Blair and John D. Steinbruner, *The Effects of Warning on Strategic Stability*, Brookings Occasional Papers (Washington, D.C.: Brookings Institution, 1991).

33. Blair, *Strategic Command and Control.*

34. Ibid., p. 284.

35. Ibid., p. 286. C^3I vulnerability can work both ways. Under some scenarios of initiation of hostilities, a side making a "limited" strike against military installations may wish to preserve the opponent's C^3I to make possible negotiations before escalation to strikes against cities.

36. Kurt Gottfried and Bruce G. Blair, eds., *Crisis Stability and Nuclear War* (New York: Oxford University Press, 1988), pp. 278–79.

37. Blair and Steinbruner, "Effects of Warning," p. 38.

38. Patrick E. Tyler, "Troubling Question of Coup: Whose Finger Was on Soviet Nuclear Trigger?" *New York Times*, Aug. 24, 1991, p. 9.

39. Abram Chayes and Jerome B. Wiesner, eds., *ABM: An Evaluation of the Decision to Deploy an Anti-Ballistic Missile System* (New York: Harper & Row, 1969); Fred Kaplan, *The Wizards of Armageddon* (New York: Simon & Schuster, 1983); Benson Adams, *Ballistic Missile Defense* (New York: American Elsevier, 1971).

40. John Newhouse, *War and Peace in the Nuclear Age* (New York: Alfred A. Knopf, 1989).

41. McGeorge Bundy, "To Cap the Volcano," *Foreign Affairs* 48, no. 1 (Oct. 1969): 10.

42. There have by now been innumerable studies of missile-defense possibili-

ties, particularly in relation to the Strategic Defense Initiative. See especially *Weapons in Space*, vols. 1, 2, *Daedalus* 114, nos. 2, 3 (Spring, Summer 1985); Office of Technology Assessment, *SDI: Technology Survivability and Software* (Washington, D.C.: OTA, May, 1988); Ashton B. Carter, *Directed Energy Missile Defense in Space*, Office of Technology Assessment Background Paper (Washington, D.C.: OTA, 1984); Sidney D. Drell, Philip J. Farley, and David Holloway, *The Reagan Strategic Defense Initiative: A Technical, Political, and Arms Control Assessment* (Palo Alto, Calif.: Stanford University Press, 1985); Joseph S. Nye, Jr., and James A. Schear, eds., *On the Defensive? The Future of SDI*, Aspen Strategy Group (Lanham, Md.: University Press of America, 1988); and Harold Brown, "Is SDI Technically Feasible?" *Foreign Affairs* 64, no. 3 (*America and the World, 1985*): 435–54.

43. Ronald Reagan, "Peace and National Security" address to the nation, Mar. 23, 1983, reprinted in part in *Weapons in Space* 2:369. A key promoter of the idea of the Strategic Defense Initiative to President Reagan was the physicist Edward Teller, a major and highly controversial figure in the postwar development of American nuclear weapons. See the summary of his role by William J. Broad (*Teller's War: The Top-Secret Story behind the Star Wars Deception* [New York: Simon & Schuster, 1992]).

44. *Weapons in Space;* Carter, *Missile Defense;* Brown, "Is SDI Technically Feasible?"; American Association for the Advancement of Science, Intersociety Working Group, *AAAS Report XIII: Research and Development, FY 1989* (Washington, D.C., 1988), p. 116; American Association for the Advancement of Science, Intersociety Working Group, *AAAS Report XIV: Research and Development, FY 1990* (Washington, D.C., 1989), p. 108.

45. The mistake in the fabrication of the Hubble telescope, designed and built under painstaking, high-quality supervision in a benign (nonhostile) environment but never fully tested before launch, demonstrates the extreme difficulty of avoiding all errors in complex systems (David Lindley, "A Discovery and New Puzzles," *Nature* 346 [Aug. 16, 1990]: 599; William J. Broad, "Measuring Device Identified as Probable Hubble Error," *New York Times*, Sept. 14, 1990, p. A9).

46. "Transcript of Group Interview with President at White House," *New York Times*, Mar. 30, 1983, p. 14; Ronald Reagan, "Reykjavik Summit: The American View," in Steven W. Guerrier and Wayne C. Thompson, eds., *Perspectives on Strategic Defense* (Boulder, Colo.: Westview Press, 1987), p. 252; Douglas Waller, James T. Bruce, and Douglas M. Cook, *The Strategic Defense Initiative: Progress and Challenges* (Claremont, Calif.: Regina Books, 1987).

47. Richard Rhodes, *The Making of the Atomic Bomb* (New York: Simon & Schuster, 1986), p. 296; "Fateful Discovery Almost Forgotten," *Nature* 337 (Feb. 9, 1989): 499–502.

48. John M. Logsdon, *Decision to Go to the Moon* (Cambridge, Mass.: MIT Press, 1970).

49. George Hutchinson, "Software Aspects of SDI," in John Holdren and Joseph Rotblat, eds., *Strategic Defenses and the Future of the Arms Race: A Pugwash Symposium* (New York: St. Martin's Press, 1987), pp. 92–95; Sanford Lakoff and Herbert F. York, *A Shield in Space? Technology, Politics, and the Strategic Defense Initiative* (Berkeley: University of California Press, 1989).

50. Reagan, "Peace and National Security," p. 370.

51. Carter, *Missile Defense*.

52. Bruce Parrott, "Soviet Policy toward BMD and SDI," in Harold Brown, ed., *The Strategic Defense Initiative: Shield or Snare* (Boulder, Colo.: Westview Press, 1987), pp. 195–231; Jeffrey Boutwell and F. A. Long, "SDI and U.S. Security," in *Weapons in Space* 2:315–30.

53. The question of the competence of Soviet r/d is discussed in chapter 4.

54. Kosta Tsipis, "A Weapon without a Purpose," *Technology Review* 94, no. 8 (Nov./Dec. 1991): 53–59; Spurgeon M. Keeny, Jr., "Press Briefing: START: The End Game and SDI," *Arms Control Today* 21, no. 7 (Sept. 1991): 5–6.

55. Matthew Bunn, "Star Wars Redux: Limited Defenses, Unlimited Dilemmas," *Arms Control Today* 21, no. 4 (May 1991): 12–18; Harold Brown, "Yes on Patriot, No on SDI," *Washington Post National Weekly Edition*, Apr. 1–7, 1991, p. 29. MIT professor Theodore A. Postol has cast serious doubt on the acclaimed performance of the Patriot in the Gulf War ("Lessons of the Gulf War Experience with Patriot," *International Security* 16, no. 3 [Winter 1991/92]: 119–71).

56. Bruce W. MacDonald, "Falling Star: SDI's Troubled Seventh Year," *Arms Control Today* 20, no. 7 (Sept. 1990): 7–11.

57. "Conventional" implies here all weapons except those that are nuclear, biological, or chemical. Many of these conventional weapons were developed originally in the context of the U.S.–USSR strategic confrontation, but are also highly relevant to non-nuclear conflict.

58. Funding for strategic programs in the U.S. defense budgets since the early 1960s has typically been only some 10 percent of the total (Benjamin J. Schemmer, "New Administration Faces Fallout over Strategic Priorities," *Armed Forces Journal International*, Nov. 1988, p. 80).

59. Paul Dickson, *The Electronic Battlefield* (Bloomington: Indiana University Press, 1976).

60. Jacques S. Gansler, *Affording Defense* (Cambridge, Mass.: MIT Press, 1989), p. 91.

61. Seymour Deitchman, *Military Power and the Advance of Technology: General Purpose Military Forces for the 1980s and Beyond* (Boulder, Colo.: Westview Press, 1979); William J. Perry, "Desert Storm and Deterrence," *Foreign Affairs* 70, no. 4 (Fall 1991): 66–82.

62. Max Hastings and Simon Jenkins describe the new situation well: "It would be difficult to overstate the impact of the *Sheffield's* loss upon the British task force. Officers and men alike were appalled, shocked, subdued by the ease with which a single enemy aircraft firing a cheap—£300,000—by no means ultra-modern sea-skimming missile had destroyed a British war ship specifically designed and tasked for air defense" (*The Battle for the Falklands* [New York: W. W. Norton, 1983], p. 155).

63. "Stung," *The Economist*, July 4, 1987, p. 41.

64. A typical piece was Lawrence Freedman, "Baghdad Stunned by Science," *The Independent*, Jan. 18, 1991, p. 21. The special conditions that helped to make the victory so devastating and so rapid are often forgotten, however: five months to mobilize forces in the area, adequate and protected nearby port facilities, an isolated enemy cut off from resupply, and terrain that maximized the effective-

ness of air power (Eric Schmitt, "Pentagon Cites Some War Lessons," *New York Times*, July 17, 1991, p. A3).

65. Evelyn Richards, "From Smart Bombs to Brilliant Missiles," *Washington Post National Weekly Edition*, Mar. 11–17, 1991, p. 34; John Lancaster, "Advancing on High-Tech Weaponry," *Washington Post National Weekly Edition*, Jan. 6–12, 1992, p. 31.

66. Michael Wines, "Third World Seeks Advanced Arms," *New York Times*, Mar. 26, 1991, p. A12.

67. Perry, "Desert Storm and Deterrence," pp. 78–79.

68. Patrick E. Tyler, "Pentagon Reassesses the Vulnerability of Its Tanks," *New York Times*, Aug. 15, 1991, p. A10.

69. Thomas L. Friedman, "Today's Threat to Peace Is the Guy down the Street," *New York Times*, June 2, 1991, p. E3.

70. Keith Krause, *The International Trade in Arms*, Background Paper no. 28, Canadian Institute for International Peace and Security (Ottawa, Mar. 1989), p. 9; Aaron Karp, "The Frantic Third World Quest for Ballistic Missiles," *Bulletin of the Atomic Scientists* 44, no. 5 (June 1988): 14; Andrew L. Ross, "Do-It-Yourself Weaponry," *Bulletin of the Atomic Scientists* 46, no. 4 (May 1990): 20–22. Ross estimated arms manufacture in developing countries rose from $2 million in 1950 to $1.1 billion (in constant 1975 dollars) in 1984.

71. Office of Technology Assessment, *Global Arms Trade: Commerce in Advanced Military Technology and Weapons*, Report Brief (Washington, D.C.: OTA, June 1991). The six are India, Egypt, Indonesia, South Korea, Taiwan, and Brazil. North Korea is now also accused of exporting technologically sophisticated armaments—including ballistic missiles and the plants to make them—to the Middle East (Elain Sciolini, "U.S. Tracks a Korean Ship Taking Missiles to Syria," *New York Times*, Feb. 21, 1992, p. A9).

72. Thomas L. McNaugher, *Ballistic Missiles and Chemical Weapons: The Legacy of the Iran-Iraq War*, Brookings General Series Reprint (Washington, D.C.: Brookings Institution, 1990), reprinted from *International Security* 15, no. 2 (Fall 1990); Michael Gordon, "Chinese Reported to Weigh Sale of Poison Gas Chemicals to Libya," *New York Times*, June 7, 1990, p. A14. Iraq has acknowledged a CW capability, which it used in the war against Iran and against its own population (McNaugher, *Ballistic Missiles*, p. 17). The possible use of CW warheads on Scud missiles fired at Israel was one of the greatest fears of the Gulf War.

73. For a comprehensive summary of arms-trade issues, see Andrew J. Pierre, *The Global Politics of Arms Sales* (Princeton: Princeton University Press, 1982). For more-recent analyses, see Office of Technology Assessment, *Global Arms Trade; Technology Review* 93, no. 4 (May/June 1990), which contains several useful articles on the subject; and Michael Brzoska and Thomas Ohlson, eds., *Arms Transfers to the Third World, 1971–1985*, Stockholm International Peace Research Institute (Oxford: Oxford University Press, 1987).

74. David B. Ottaway, "Mules, Missiles, and Mujaheddin," *Washington Post National Weekly Edition*, Feb. 20–26, 1989, p. 6. The wonderful irony is that the high-technology Stingers had to be carried into Afghanistan on the backs of pack mules.

75. David B. Ottaway, "Surprise! The Afghan Rebels Can Fire a Stinger," *Washington Post National Weekly Edition*, Aug. 3, 1987, pp. 17–18.

76. Office of Technology Assessment, *Global Arms Trade*, p. 1.

77. U.S. Arms Control and Disarmament Agency, *World Military Expenditures and Arms Transfers, 1989* (Washington, D.C.: ACDA, 1990), p. 11. Russia has said it must sell arms for economic reasons ("Russians Decide to Sell More Arms," *Boston Globe*, Feb. 23, 1992, p. 2).

78. "Praise God and Hold the Ammunition," *The Economist*, June 8, 1991, p. 16; Eric Schmitt, "Cheney Says U.S. Plans New Arms Sales to the Mideast," *New York Times*, June 5, 1991, p. A3; William D. Hartung, "Relighting the Mideast Fuse: U.S. Policy Is Big Talk—and Big Deals," *New York Times*, Sept. 20, 1991, p. A27; William Hartung, "The Boom at the Arms Bazaar," *Bulletin of the Atomic Scientists* 47, no. 8 (Oct. 1991): 14–20; Barry Schweid, "$6 Billion in US Arms Sent to Mideast since May," *Boston Globe*, Feb. 15, 1992, p. 3.

79. Steven Greenhouse, "U.S. and Allies Move to Ease Cold War Limits on Exports," *New York Times*, May 25, 1991, p. 1; Eliot Marshall, "War with Iraq Spurs New Export Controls," *Science* 251 (Feb. 1, 1991): 512; Michael Mastanduno, "The United States Defiant: Export Controls in the Postwar Era," *Daedalus* 120, no. 4 (Fall 1991): 91–112.

80. Craig R. Whitney, "U.S. and 4 Other Big Arms Makers Adopt Guidelines on Sales," *New York Times*, Oct. 20, 1991, p. 11.

81. The Missile Technology Control Regime (MTCR) includes the G-7 countries (Canada, France, Germany, Italy, Japan, the United Kingdom, and the United States, plus the European Community). Several others, including the Soviet Union (before it disappeared), have pledged to adhere to its restrictions ("Five Steps toward Regional Arms Control," *F.A.S. Public Interest Report* [Federation of American Scientists, Washington, D.C.], Mar./Apr. 1991, p. 3). For a succinct analysis of ballistic-missile proliferation, see Janne E. Nolan, "The Politics of Proliferation," *Issues in Science and Technology* 8, no. 1 (Fall 1991): 63–69.

82. Randall Forsberg, "The Other Arms-Cut Treaty," *Boston Globe*, July 20, 1991, p. 19; "Powell Sees NATO Holding Edge in Arms-Limitation Treaty," *Boston Globe*, July 17, 1991, p. 12.

83. James E. Goodby, "A New European Concert: Settling Disputes in CSCE" and Harald Mueller, "A United Nations of Europe and North America," *Arms Control Today* 21, no. 1 (Jan./Feb. 1991): 3–8; "European Security Group Grows Along with Debate over Its Role," *International Herald Tribune*, Jan. 31, 1992, p. 5.

84. Trevor Rowe, "Tough UN Plan Proposed to Monitor Iraqi Weapons," *Boston Globe*, Aug. 3, 1991, p. 2; Jerry Gray, "U.N. Using U.S. Spy Planes to Monitor Iraqi Arms," *New York Times*, Aug. 13, 1991, p. A5.

85. Joseph S. Nye, Jr., "Nonproliferation: A Long-Term Strategy," *Foreign Affairs* 56, no. 3 (Apr. 1978): 601–23.

86. Gerard C. Smith and Helena Cobban, "A Blind Eye to Nuclear Proliferation," *Foreign Affairs* 68, no. 3 (1989): 53–70; Joseph F. Pilat and Robert E. Pendley, eds., *Beyond 1995: The Future of the NPT Regime* (New York: Plenum Press, 1990).

87. Joseph A. Yager, *International Cooperation in Nuclear Energy* (Washington, D.C.: Brookings Institution, 1981), pp. 25–40; Lawrence Scheinman, *The Nonproliferation Role of the International Atomic Energy Agency*, Resources for the Future (Baltimore: Johns Hopkins University Press, 1985).

88. France, China, and South Africa have signaled an intention to join the NPT ("South Africa to Sign," *Nature* 352 [July 4, 1991]: 7; Jon B. Wolfsthal, "China Nears NPT Membership, U.S. to Drop Missile Sanctions," *Arms Control Today* 22, no. 1 [Jan./Feb. 1992]: 46). Seymour Hersh estimates Israel has about three hundred nuclear bombs (*The Samson Option: Israel, America, and the Bomb* [New York: Random House, 1991]). Signing the treaty does not ensure that a nation will not seek to develop nuclear weapons, as the examples of Iraq and North Korea illustrate (Jeff Gerth, "6 Held in Scheme to Send Atom Gear to Iraq," *New York Times*, Mar. 29, 1990, p. 1; Andrew Mack, "North Korea and the Bomb," *Foreign Policy*, no. 83 [Summer 1991]: 87–104; Elaine Sciolino, "C.I.A. Chief Says North Koreans Plan to Make Secret Atom Arms," *New York Times*, Feb. 26, 1992, p. 1).

89. For a discussion of the political disincentives for acquiring a nuclear capability, see Stephen M. Meyer, *The Dynamics of Nuclear Proliferation* (Chicago: University of Chicago Press, 1984): on the situation after the end of the cold war, see George W. Rathjens and Marvin Miller, "Nuclear Proliferation after the Cold War," *Technology Review* 94, no. 6 (Aug./Sept. 1991): 25–32.

90. Gerth, "6 Held in Scheme"; Stephen Engelberg, "CIA Says Pakistan Is Working to Gain Nuclear 'Capability'," *International Herald Tribune*, May 20–21, 1989, p. 3; Gerald M. Steinberg, "The Middle East in the Missile Age," *Issues in Science and Technology* 5, no. 4 (Summer 1989): 35–40; Ricardo Bonalume, "Democratic Brazil Says No," *Nature* 347 (Oct. 4, 1990): 417; Mack, "North Korea and the Bomb."

91. William J. Broad, "Iraqi Atom Effort Exposes Weakness in World Controls," *New York Times*, July 15, 1991, p. 1; R. Jeffrey Smith, "Grim Lessons for the World in Iraq's Secret Nuclear Program," *Washington Post National Weekly Edition*, Aug. 19–25, 1991, p. 22; David Albright and Mark Hibbs, "Iraq's Nuclear Hide-and-Seek," *Bulletin of the Atomic Scientists* 47, no. 7 (Sept. 1991): 14–23; Elaine Sciolini, "Iraq's Nuclear Program Shows the Holes in U.S. Intelligence," *New York Times*, Oct. 20, 1991, p. 5.

92. George N. Grammas, "Multilateral Responses to the Iraqi Invasion of Kuwait: Economic Sanctions and Emerging Proliferation Controls," *Maryland Journal of International Law and Trade* 15, no. 1 (Spring 1991): 1–21; William J. Broad, "New Atom Accord Hailed by Experts," *New York Times*, May 3, 1992, p. 15.

93. For an excellent summary of the scale and nature of the problem of the former Soviet Union's nuclear arsenal and what might be done about it, see Kurt M. Campbell et al., *Soviet Nuclear Fission: Control of the Nuclear Arsenal in a Disintegrating Soviet Union*, CSIA Studies in International Security No. 1, Center for Science and International Affairs, John F. Kennedy School of Government, Harvard University, Cambridge, Mass., Nov. 1991.

94. Nuclear Energy Policy Study Group, *Nuclear Power Issues and Choices* (Cambridge, Mass.: Ballinger, 1977).

95. Stansfield Turner and Thomas Davies, "Plutonium Terror on the High Seas," *New York Times*, Apr. 28, 1990, p. 25; David Swinbanks, "Japan Debates Plutonium," *Nature* 352 (July 4, 1991): 7; Tatsujiro Suzuki, "Japan's Nuclear Dilemma," *Technology Review* 94, no. 7 (Oct. 1991): 42–49.

96. Turner and Davies, "Plutonium Terror."

97. Ben Sanders and John Simpson, *Nuclear Submarines and Non-Proliferation: Cause for Concern*, Occasional Paper Number Two, Program for Promoting Nuclear Non-Proliferation, Center for International Policy Studies, University of Southampton, Southampton, England, 1988.

98. Japan is not wholly without military forces; its "self-defense force" is a substantial commitment. Even though it represents expenditures of only about 1 percent of GNP, the country's economic size is now so great that even that 1 percent makes Japan's defense sector among the world's largest in absolute terms (Michael W. Chinworth, "Japan's Dilemma: Rising Defense Budgets in an Era of Global Cuts," *Breakthroughs* [MIT Program on Defense and Arms Control Studies, Cambridge, Mass.], Fall 1990). Voices have begun to be raised in Japan questioning the restrictions on its military capabilities (Tetsuya Kataoka, *Waiting for a "Pearl Harbor": Japan Debates Defense* [Stanford, Calif.: Hoover Institution Press, 1980]).

99. Osgood Caruthers, "Soviet Downs American Plane; U.S. Says It Was Weather Craft; Kruschev Sees Summit Blow," *New York Times*, May 6, 1960, p. 1.

100. Wladyslaw Kozaczuk, *Enigma: How the German Machine Cipher Was Broken, and How It Was Used by the Allies in World War Two*, ed. and trans. Christopher Kasparek (Frederick, Md.: University Publications of America, 1984; F. W. Winterbotham, *The Ultra Secret* (New York: Harper & Row, 1974).

101. Bruce D. Berkowitz and Allan E. Goodman, *Strategic Intelligence for American National Security* (Princeton: Princeton University Press, 1989), pp. 64–84.

102. Kenneth Flamm, *Targeting the Computer: Government Support and International Competition* (Washington, D.C.: Brookings Institution, 1987), pp. 49–51; Alic et al., *Beyond Spinoff*.

103. David Kahn, "Cryptology Goes Public," *Foreign Affairs* 58, no. 1 (Fall 1979): 142–59; Walter Sullivan, "Tighter Security Rules for Advances in Cryptology," *New York Times*, June 1, 1981, p. 14.

104. One result of the surprise of the rapid disintegration in Eastern Europe and the Soviet Union was a call for greater emphasis on human rather than technological means for obtaining intelligence information (Michael Wines, "New-Era Espionage," *New York Times*, Sept. 17, 1991, p. A13).

105. Sciolini, "Iraq's Nuclear Program."

106. Stansfield Turner, "Intelligence for a New World Order," *Foreign Affairs* 70, no. 4 (Fall 1991): 150–66.

107. Colum Lynch, "Postwar Spy Business Is Booming," *Boston Globe*, Oct. 25, 1991, p. 3.

108. Turner, "Intelligence," p. 152.

109. The negotiations over the nuclear-test ban starting in 1958 were the first major attempt after World War II to find a way to lessen the danger of the arms race, though the effects of radioactive fallout from tests in the atmosphere played

a large part in stimulating the negotiations (National Academy of Sciences, Committee on International Security and Arms Control, *Nuclear Arms Control: Background and Issues* [Washington, D.C.: National Academy Press, 1985]).

110. *Report of Panel on Nuclear Weapons Safety of the House Armed Services Committee, Dec. 1990*, excerpted in "How Safe Is Safe," *Bulletin of the Atomic Scientists* 47, no. 3 (Apr. 1991): 35–40. The argument is sometimes made that the more weapons there are, the more likely it is they will actually be fired in anger. But if there were a rough equivalence of weapons on each side, in numbers either much larger or even considerably smaller than the number presently deployed, the risk of unacceptably destructive retaliation, which would inhibit intentional use, would be essentially the same. The danger of *accidental* discharge of nuclear weapons, however, increases as the numbers of weapons increases. The probability that a given weapon will be fired by accident by a major nuclear power may be trivially small, but it cannot be reduced to zero without making the weapons unusable when wanted. The net probability of accident for all the weapons therefore increases with each added weapon; and if there is any possibility of common-mode (interactive) failures, the probability of accident becomes even greater. In reality, therefore, the more weapons in existence, the higher the chance—even if small—that one will be discharged by a technical or human mistake.

111. At the Reykjavik summit in 1987, Presidents Reagan and Gorbachev came close to an agreement to eliminate nuclear weapons, though the agreement would probably not have been ratified (James Schlesinger, "Reykjavik and Revelations: A Turn of the Tide," *Foreign Affairs* 65, no. 3 [*America and the World, 1986*]: 426–46).

112. Eric Schmitt, "Senate Approval and Sharp Debate Seen," *New York Times*, July 19, 1991, p. A7.

113. Office of Technology Assessment, *Verification Technologies: Cooperative Aerial Surveillance in International Agreements* (Washington, D.C.: OTA, July 1991); Kenneth Adelman, Antonia Handler Chayes and Abram Chayes, Lewis A. Dunn, and Ivan Oerlich, "Policy Focus: Arms Control Verification Reconsidered," *International Security* 14, no. 4 (Spring 1990): 140–84.

114. Abraham D. Sofaer, "The ABM Treaty and the SDI," *Harvard Law Review* 99, no. 8 (June 1986): 1972–85; Matthew Bunn, *Foundation for the Future: The ABM Treaty and National Security* (Washington, D.C.: Arms Control Association, 1990), pp. 58–73.

115. George W. Rathjens, Abram Chayes, and Jack Ruina, *Nuclear Arms Control Agreements* (Washington, D.C.: Carnegie Endowment for International Peace, 1974), pp. 15–17.

116. Ibid., pp. 15–16.

117. Harvey Brooks, "The Military Innovation System and the Qualitative Arms Race," *Daedalus* 104, no. 3 (Summer 1975): 75–98. A fascinating, though not well known, example of an attempt to prevent a weapons deployment through an agreed ban on testing was the proposal of Vannevar Bush in 1952 to proceed with the development of the H-bomb but to refrain from testing it while seeking agreement with the Soviets for similar restraint. Both sides would then be confident that they had the technology and that violation by either side would be immediately detected. The H-bomb was too complex and untried to be able to be deployed

without prior testing. The proposal received little support (Hans Bethe, "Sakharov's H-bomb," *Bulletin of the Atomic Scientists* 46, no. 8 [Oct. 1990]: 9).

118. Trevor Findlay, *Nuclear Dynamite: The Peaceful Nuclear Explosions Fiasco* (Sydney: Brassey's Australia, 1990).

119. Bunn, *Foundation for the Future*, pp. 58–73.

120. An example of the typical resistance encountered was the assertion by the U.S. Department of Energy in the spring of 1990, after years of consideration of a comprehensive nuclear-test ban, that it would take ten years to "evaluate whether we can maintain the U.S. nuclear deterrent in the event of further test limitations" ("Continued Testing 'Essential,' DOE Tells Congress," *Arms Control Today* 20, no. 4 [May 1990]: 29).

121. The 1991 START agreement ran to some 750 pages and represented only a minor—though symbolically important—actual reduction in weapons and warheads (R. Jeffrey Smith, "Beating Swords into Trade Agreements," *Washington Post National Weekly Edition*, Aug. 5–11, 1991, p. 16).

122. Richard G. Hewlett and Oscar E. Anderson, Jr., *The New World, 1939/ 1946*, vol. 1 of *A History of the United States Atomic Energy Commission* (University Park, Pa.: Pennsylvania State University Press, 1962); Richard G. Hewlett and Francis Duncan, *Atomic Shield, 1947/1952*, vol. 2 of *A History of the United States Atomic Energy Commission* (University Park, Pa.: Pennsylvania State University Press, 1969).

123. Hewlett and Anderson, *The New World*, pp. 534–54.

124. Ibid., pp. 531–619.

125. Iraq's ability to reach Israel with Scud missiles in the Gulf War very nearly brought that nation into the conflict, which would have likely split the allied coalition. The United States went to great lengths to keep Israel from retaliating.

126. The seven major Western nations called for an arms-transfer registry at their 1991 economic summit conference (R. W. Apple, Jr., "7 Nations Voice Support for Gorbachev," *New York Times*, July 17, 1991, p. 1).

127. Stuart Auerbach, "The U.S. Achilles' Heel in Desert Storm," *Washington Post National Weekly Edition*, Apr. 1–7, 1991, p. 11. Dependence is discussed more generally in the next chapter.

Chapter Four

1. One set of issues, those related to apparent dangers for the entire planet, will be deferred until chapter 5. They can more readily be understood independently of others, and may have special significance for the structure of the international system.

2. Computer prices have fallen, on average, more than 20 percent per year, a decrease eclipsing any sustained price decline in recorded economic history (Kenneth Flamm, *Creating the Computer: Government, Industry and High Technology* [Washington, D.C.: Brookings Institution, 1988], p. 1). The number of components per silicon chip—the core of the integrated circuits essential for the high-capability, low-cost computer—is still increasing by a factor of one hundred per decade, and the capacity of light-wave communications, measured by the capacity of a circuit times the distance, is doubling every year (John S. Mayo, "The

Evolution of Information Technologies," in Bruce R. Guile, ed., *Information Technologies and Social Transformation* [Washington, D.C.: National Academy Press, 1985], pp. 11, 18). Growth in the software market from 1985 to 1991 was expected to be 16–20 percent per year in the United States (Lydia Arossa, "Software and Computer Services," *OECD Observer*, Apr./May 1988, pp. 13–16).

3. There is an extensive literature on this subject. For an excellent, brief article that draws parallels with the industrial revolution, see Melvin Kranzberg, "The Information Age: Evolution or Revolution?" in Guile, *Information Technologies*, pp. 35–53. Some useful compendiums are Ithiel de Sola Pool, *Technologies without Boundaries: On Telecommunications in a Global Age*, ed. Eli M. Noam (Cambridge, Mass.: Harvard University Press, 1990); Oswald H. Ganley and Gladys D. Ganley, *To Inform or to Control? The New Communications Networks*, 2d ed. (Norwood, N.J.: Ablex, 1989); and Arnold Pacey, *Technology in World Civilization: A Thousand-Year History* (Cambridge, Mass.: MIT Press, 1990).

4. James Brian Quinn, "The Impacts of Technology in the Services Sector," in Guile, *Information Technologies*, p. 119.

5. W. Michael Blumenthal, "The World Economy and Technological Change," *Foreign Affairs* 66, no. 3 (1987/88): 534.

6. Office of Technology Assessment, *Technology and the American Economic Transition: Choices for the Future*, Summary (Washington, D.C.: OTA, May 1988), p. 16.

7. James N. Rosenau, *Turbulence in World Politics: A Theory of Change and Continuity* (Princeton: Princeton University Press, 1990) Rosenau's argument is discussed in chapter 7.

8. The literatures in interdependence, dependencia, and international regimes are particularly relevant. Most analyses deal only cursorily with how continuing technological change actually brings about or interacts with the political phenomena under study. There are many texts and references in these subjects; those especially valuable for the quality of their analyses and for their comprehensive references to other works are Robert O. Keohane and Joseph S. Nye, Jr., *Power and Interdependence: World Politics in Transition* (Boston: Little, Brown, 1977); Stephen D. Krasner, ed., *International Regimes* (Ithaca, N.Y.: Cornell University Press, 1983; and Fernando Henrique Cardoso and Enzo Faletto, *Dependency and Development in Latin America* (Berkeley: University of California Press, 1979).

9. The eminent historian Michael Howard, in commenting on the changes in the Soviet Union, said, "Historians and economists will long debate the causes of these changes, but they are likely to agree about one factor: the supreme importance of the communications revolution" ("The Remaking of Europe," prepared text for Alastair Buchan Memorial Lecture, May 12, 1990, reprinted in *Survival* 32, no. 2 [Mar./Apr. 1990]: 100).

10. George H. Quester, *The International Politics of Television* (Lexington, Mass.: Lexington Books, 1990), pp. 125–53; Elie Abel, *The Shattered Bloc: Behind the Upheaval in Eastern Europe* (Boston: Houghton Mifflin, 1990), pp. 126, 130, 135.

11. In the former German Democratic Republic, environmental groups collected information on environmental degradation (classified as secret by the regime) and passed it on to Western television. For the first time, citizens of that

country, watching West German television, could see the countrywide conditions in which they lived (talk by Carlo Jordan of the Green Network, East Berlin, at a conference at the Harvard Center for European Studies, Apr. 7, 1991).

12. "If television falls, the revolution falls," said Eugenia Bogdan of Romanian television a few days after the arrest and subsequent execution of dictator Nicolae Ceausescu (*New York Times*, Dec. 28, 1989, p. 1). In an astonishing note that same day, the *New York Times* reported that the Jews in Romania were safe, the information obtained from a handwritten fax message from a Romanian rabbi to a rabbi in New York (p. 12).

13. Michael Mandelbaum, "Coup de Grace: The End of the Soviet Union," *Foreign Affairs* 71, no. 1 (*America and the World, 1991/92*): 164–83.

14. Robert A. Scalapino, "China: What Leaders for What Policy?" *International Herald Tribune*, June 26, 1989, p. 4.

15. Ibid.

16. Nicholas K. Kristof, "China Bans Pro-Student Newspaper," *New York Times*, Apr. 25, 1989, p. 3; idem, "Beijing Hints at Crackdown on Students," *New York Times*, Apr. 26, 1989, p. 3; idem, "Beijing Dismisses Outspoken Editor," *New York Times*, Apr. 27, 1989, p. 7; Constance L. Hays, "China Orders End to TV Broadcasts," *New York Times*, May 20, 1989, p. 7; E. J. Dionne, Jr., "TV Steps into the Fray and Alters It," *New York Times*, May 21, 1989, p. 18.

17. For extended analyses of the role of information technologies, see Ithiel de Sola Pool, *Technologies of Freedom* (Cambridge, Mass.: Harvard University Press, Belknap Press, 1983); Pool, *Technologies without Boundaries*; Karl W. Deutsch, *Nationalism and Social Communication: An Inquiry into the Foundations of Nationalism* 2d ed. (Cambridge, Mass.: MIT Press, 1966); Majid Tehranian, *Technologies of Power: Information Machines and Democratic Prospects* (Norwood N.J.: Ablex, 1990); Iain McLean, *Democracy and New Technology* (Oxford: Blackwell, 1989); and Seymour Goodman, "Information Technologies and the Citizen: Toward a 'Soviet-Style Information Society'?" in Loren R. Graham, ed., *Science and the Soviet Social Order* (Cambridge, Mass.: Harvard University Press, 1990), pp. 51–70.

18. Scalapino, "China"; Richard Bernstein, "To Old Hands China Feels Distressingly Familiar," *International Herald Tribune*, June 21, 1989, p. 6. The attempt to control the movement of information continues: in 1990, the Chinese government revised its telephone system to prevent certain internal communications and to restrict access by foreigners (Sheryl WuDunn, "China Is Blocking Phone Contacts," *New York Times*, Feb. 24, 1991, p. 15).

19. Willi A. Boelcke, ed., *The Secret Conferences of Dr. Goebbels: The Nazi Propaganda War, 1939–45*, trans. Ewald Osers (New York: E. P. Dutton, 1970); Ernst Kris and Hans Speier, *German Radio Propaganda* (New York: Oxford University Press, 1944).

20. George Orwell's *1984* has probably had the greatest impact on the perception of the political dangers of technologies used for authoritarian control.

21. S. Frederick Starr, "Soviet Union: A Civil Society," *Foreign Policy*, no. 70 (Spring 1988): 26–41; idem, "New Communications Technology and Civic Culture in the USSR," in Graham, *Science and the Soviet Social Order*, pp. 19–50.

22. Starr, "New Communications Technology," pp. 26–30; Karl W. Deutsch,

Political Community and the North Atlantic Area (Princeton: Princeton University Press, 1957), p. 51.

23. Starr, "New Communications Technology," p. 41.

24. Ibid., p. 42.

25. Ibid., p. 50.

26. Starr, "Soviet Union: A Civil Society," p. 35.

27. A more cautious view is presented in Goodman, "Information Technologies and the Citizen."

28. U.S. Department of State, Bureau of Intelligence and Research, *Soviet Influence Activities: A Report on Active Measures and Propaganda, 1987–88* (Washington, D.C., Aug. 1989); Ladislav Bittman, *The KGB and Soviet Disinformation: An Insider's View* (Washington, D.C.: Pergamon-Brassey's, 1985); Ladislav Bittman, ed., *The New Image-Makers: Soviet Propaganda and Disinformation Today* (Washington, D.C.: Pergamon-Brassey's, 1988); Boelcke, *Secret Conferences.*

29. Many of the chapters in a 1990 review of international affairs assembled by the Council on Foreign Relations refer to the underlying effects of the information revolution as an essential part of the explanation for a "sea-change" in world affairs (Nicholas X. Rizopoulos, ed., *Sea-Changes: American Foreign Policy in a World Transformed* [New York: Council on Foreign Relations, 1990]).

30. Stanley Hoffmann, "A New World and Its Troubles," in Rizopoulos, *Sea-Changes,* p. 280.

31. Joan E. Spero, "Guiding Global Finance," *Foreign Policy,* no. 73 (Winter 1988/89): 114–15. This article summarizes well much other, more technical, literature.

32. John Burgess, "The Market That Never Sleeps," *Washington Post National Weekly Edition,* Nov. 16, 1987, p. 7.

33. C. Michael Aho and Marc Levinson, "The Economy After Reagan," *Foreign Affairs* 67, no. 2 (Winter 1988/89): 18.

34. Joel A. Bleeke and Lowell L. Bryan, "The Globalization of Financial Markets," *McKinsey Quarterly,* Winter 1988, p. 25.

35. Spero, "Guiding Global Finance," p. 119.

36. Colin Mayer, "The Changing Structure of Financial Markets," *Amex Bank Review,* Special Papers no. 11, Feb. 1987, p. 3.

37. Thomas D. Steiner and Diogo Teixeira, "Technology Is More than Just a Strategy," *McKinsey Quarterly,* Winter 1988, p. 39.

38. Kurt Eichenwald and John Markoff, "Wall Street's Souped Up Computers," *New York Times,* Oct. 16, 1988, Business section, p. 1.

39. "High Hopes, High Costs for Wall St.'s High Technology," *The Economist,* Feb. 2, 1991, p. 75.

40. Several exchanges are already on a twenty-four-hour basis or moving toward it, to allow worldwide trading at any hour (Stephen Labaton, "Extra Time for Trade Expected," *New York Times,* May 20, 1991, p. D1; idem, "S.E.C. Supports Plan for Earlier Trading of Stocks in the U.S.," *New York Times,* Oct. 11, 1991, p. 1). However, brokers are objecting to the longer hours; the New York Stock Exchange backed away temporarily from its first steps to expand hours (Floyd Norris, "Big Board Won't Add Hours," *New York Times,* July 30, 1991, p. D1).

41. Spero, "Guiding Global Finance," p. 116.

42. Ibid., p. 118.

43. Ibid., p. 120.

44. Miles Kahler offers a typical comment: "For national governments, . . . the effects [of financial integration] are not obscure: their autonomy in setting economic policies is increasingly constrained, as capital controls become less effective and financial markets read their political and economic missteps with shorter and shorter time lags" ("The International Political Economy," in Rizopoulos, *Sea-Changes*, p. 102).

45. Aho and Levinson, "The Economy after Reagan," p. 18; Rudiger Dornbusch and Jeffrey Frankel, "The Flexible Exchange Rate System: Experience and Alternatives," in Silvio Borner, ed., *International Finance and Trade in a Polycentric World* (New York: Macmillan International Press Association, 1988), pp. 151–97.

46. Robert Sobel, *Panic on Wall Street: A Classic History of America's Financial Disasters with a New Exploration of the Crash of 1987* (New York: Truman Talley Books, 1988); "Excerpts from 'Brady: The Report of the Presidential Task Force on Market Mechanisms,'" in Robert W. Kamphius, Jr., et al., eds., *Black Monday and the Future of the Financial Markets* (Homewood, Ill.: Dow Jones-Irwin, 1989), pp. 127–204; Anthony D. Loehnis, "Volatility in Global Securities Markets," *Economic Insights*, Nov./Dec. 1990, pp. 16–19.

47. "Why Stockmarkets Move Together," *The Economist*, Mar. 11, 1989, p. 89.

48. Spero, "Guiding Global Finance," p. 127; Ethan B. Kapstein, "Resolving the Regulator's Dilemma: International Coordination of Banking Regulations," *International Organization* 43, no. 2 (Spring 1989): 328.

49. Diana B. Henriques, "Seeking Global Rules for Brokers," *New York Times*, July 14, 1991, p. F15; idem, "In World Markets, Loose Regulation," *New York Times*, July 23, 1991, p. D1.

50. Quoted in Spero, "Guiding Global Finance," p. 125.

51. "London's Share Markets: Outlook Unsettled," *The Economist*, May 4, 1991, p. 78.

52. Kapstein, "Resolving the Regulator's Dilemma," p. 330.

53. Ibid., p. 329.

54. Spero, "Guiding Global Finance," p. 134; W. Michael Blumenthal, "The World Economy," p. 549.

55. Richard J. Barnet and Ronald E. Müller, *Global Reach: The Power of the Multinational Corporations* (New York: Simon & Schuster, 1974), p. 363.

56. Ibid., p. 366.

57. There is an extensive literature on multinational corporations. See especially Barnet and Müller, *Global Reach;* Raymond Vernon, *Sovereignty at Bay* (New York: Basic Books, 1971); Charles P. Kindleberger, *Multinational Excursions* (Cambridge, Mass.: MIT Press, 1984); Robert Gilpin, *U.S. Power and the Multinational Corporation: The Political Economy of Foreign Direct Investment* (New York: Basic Books, 1975); and, for a brief summary of the debate, Charles W. Kegley, Jr., Eugene R. Wittkopf, and Lucia W. Rawls, "The Multinational Corporation: Curse or Cure?" in Charles W. Kegley and Eugene R. Wittkopf, eds., *The Global Agenda: Issues and Perspectives*, 2d ed. (New York: Random House, 1988),

p. 272; and David Leyton-Brown, "The Roles of the Multinational Enterprise in International Relations," in David G. Haglund and Michael K. Hawes, eds., *World Politics: Power, Interdependence, and Dependence* (Toronto: Harcourt Brace Jovanovich, 1990). For a summary of the literature arguing that the power and autonomy of the state have been enhanced, see Peter B. Evans, Dietrich Rueschemeyer, and Theda Skocpol, eds., *Bringing the State Back In* (New York: Cambridge University Press, 1985).

58. The competitiveness issue is discussed in the next section of this chapter.

59. Joseph Grunwald and Kenneth Flamm, *The Global Factory: Foreign Assembly in International Trade* (Washington, D.C.: Brookings Institution, 1985), p. 12. See also Harley Shaiken, "High Tech Goes Third World," *Technology Review* 91, no. 1 (Jan. 1988): 38–47; and Bruce Stokes, "Mexican Momentum," *National Journal* 19 (June 20, 1987): 1572–78.

60. Thomas H. Lee and Proctor P. Reid, eds., *National Interests in an Age of Global Technology* (National Academy of Engineering (Washington, D.C.: National Academy Press, 1991), p. 25.

61. Lawrence C. McQuade, "The Challenge to Business Leadership," in Charles A. Cerami, ed., *A Marshall Plan for the 1990s: An International Roundtable on World Economic Development* (New York: Praeger, 1989), p. 186.

62. Yves Doz, "International Industries: Fragmentation vs. Globalization," in Bruce R. Guile and Harvey Brooks, eds., *Technology and Global Industry: Companies and Nations in the World Economy*, National Academy of Engineering (Washington, D.C.: National Academy Press, 1987), p. 98. Much of the discussion in the text with regard to the role of technology in dispersed production is drawn from this excellent report.

63. Ibid., p. 108. Michael Porter shows in his research how U.S. companies that chose to move production from the United States to low-labor-cost countries have now been replaced in the United States by Japanese companies that substituted automation for labor ("The Competitive Advantage of Nations," *Harvard Business Review*, Mar./Apr. 1990, p. 79). A U.S. National Research Council report shows that direct labor costs in most manufacturing industries are 15 percent or less of production costs and are increasingly under 5 percent in high-technology industries (National Research Council, *The Internationalization of Manufacturing: Causes and Consequences* [Washington, D.C.: National Academy Press, 1990], p. 23). Research based on studies of 686 large, technologically active firms by Pari Patel and Keith Pavitt leads them to conclude that "the production of technology remains far from globalized . . . [and] remains highly 'domesticized'" ("Large Firms in the Production of the World's Technology: An Important Case of 'Non-Globalization'," *Journal of International Business Studies* 22, no. 1 [First Quarter 1991]: 17).

64. The direction of technology-induced change remains difficult to predict. In the 1950s, it was conventional wisdom that computers would lead to centralization of firms, with great industrial concentration. The exact opposite has occurred ("The Incredible Shrinking Company," *The Economist*, Dec. 15, 1990, p. 65).

65. Harvey Brooks and Bruce R. Guile, "Overview," in Guile and Brooks, *Technology and Global Industry*, pp. 4–5. Their thesis is corroborated in part as American companies increasingly turn from the export of goods to the establish-

ment of production facilities in target countries, not for reasons of production efficiency, but in order to better serve and capture local markets (Louis Uchitelle, "Trade Barriers and Dollar Swings Raise Appeal of Factories Abroad; American Companies Turn Away from Exports," *New York Times*, Mar. 26, 1989, p. 1).

66. Doz, "International Industries," pp. 111–14.

67. Edward M. Graham and Paul R. Krugman illustrate the point with particular reference to the automobile-quota agreement with Japan, which "distorted consumer choice, reduced competition in the domestic industry, . . . and actually transferred the quota rents to the Japanese" (*Foreign Direct Investment in the United States* [Washington, D.C.: Institute for International Economics, 1989], p. 70).

68. Simon Kuznets, *Modern Economic Growth: Rate, Structure, and Spread* (New Haven, Conn.: Yale University Press, 1966), pp. 9–10.

69. An excellent, detailed summary of economic theory on the subject and of the primary writings and issues is given in National Academy of Sciences, "A Background Review of the Relationships between Technological Innovation and the Economy," in *Technology, Trade, and the U.S. Economy*, report of a workshop held at Woods Hole, Mass., Aug. 22–31, 1976 (Washington, D.C.: National Academy of Sciences, 1978), pp. 18–48.

70. Robert M. Solow, "Technical Change and the Aggregate Production Function," *Review of Economics and Statistics* 39, no. 3 (Aug. 1957): 312–20.

71. Edwin Mansfield, "Estimates of the Social Returns from Research and Development," in Margaret O. Meredith, Stephen D. Nelson, and Albert H. Teich, eds., *AAAS Science and Technology Policy Yearbook, 1991*, Committee on Science, Engineering, and Public Policy (Washington, D.C.: American Association for the Advancement of Science, 1991), pp. 314–15. On this subject, see also Edwin Mansfield et al., *Technology Transfer, Productivity, and Economic Policy* (New York: W. W. Norton, 1982); idem, *The Economics of Technological Change* (New York: W. W. Norton, 1968); Edward Denison, *The Sources of Growth in the U.S. and the Alternatives before Us* (New York: Committee for Economic Development, 1962); idem, *Why Growth Rates Differ: Postwar Experience in Nine Western Countries* (Washington, D.C.: Brookings Institution, 1967); Nestor E. Terleckyj, ed., *The State of Science and Research: Some New Indicators* (Boulder, Colo.: Westview Press, 1977); Zvi Griliches, ed., *R&D, Patents, and Productivity* (Chicago: University of Chicago Press, 1984); Christopher Freeman, *Technology Policy and Economic Performance: Lessons from Japan*, Science Policy Research Unit, University of Sussex (New York: Pinter, 1987); and Jacob Schmookler, *Invention and Economic Growth* (Cambridge, Mass.: Harvard University Press, 1966). Also, an OTA study summarizes the state of knowledge in this subject to the time of its publication (*Research Funding as an Investment: Can We Measure the Returns?* Science Policy Study Background Report no. 12, transmitted to the House Committee on Science and Technology, Task Force on Science Policy, Dec. 1986).

72. The changes in these and other traditional geopolitical factors are discussed later in this chapter.

73. Raymond Barre, talk given at Davos, Switzerland, quoted in *International Herald Tribune*, Jan. 30, 1989.

74. John A. Young, "Technology and Competitiveness: A Key to the Economic Future of the United States," *Science* 241 (July 15, 1988): 313–16; Charles H. Ferguson, "America's High-Tech Decline," *Foreign Policy* no. 74 (Spring 1989): 123–44; N. Bruce Hannay, "On the Competitive Status of U.S. Industry," *High Technology*, Dec. 1985, pp. 11–12; Harvey Brooks, "The Technological Factor in U.S. Competitiveness," *Business in the Competitive World*, Autumn 1989, pp. 81–86.

75. John M. Alic et al., eds., *Beyond Spinoff: Military and Commercial Technologies in a Changing World* (Boston: Harvard Business School Press, 1992), chap. 1.

76. The literature on international trade is vast, specialized, and controversial. For the nonspecialist, recent works of particular value are Paul R. Krugman, ed., *Strategic Trade Policy and the New International Economics* (Cambridge, Mass.: MIT Press, 1986); Michael E. Porter, *The Competitive Advantage of Nations* (New York: Free Press, 1990); and Laura D'Andrea Tyson, *Who's Bashing Whom? Trade Conflict in High-Technology Industries* (Washington, D.C.: Institute for International Economics, forthcoming, 1992), chap. 2. This section draws on all three works.

77. Porter, "Competitive Advantage of Nations," p. 78; Tyson, *Who's Bashing Whom?* chap. 2.

78. National Science Board, *Science and Engineering Indicators, 1989*, NSB 89-1 (Washington, D.C., 1989), app., table 7-11.

79. Office of Technology Assessment, *Technology and the American Transition* (Washington, D.C.: OTA, May 1988), p. 18.

80. National Science Foundation, *International Science and Technology Data Update: 1991*, NSF 91–309 (Washington, D.C., 1991), p. 127; *OECD in Figures: Supplement to the OECD Observer* (Organization for Economic Cooperation and Development, Paris), no. 170 (June/July 1991): 56–57.

81. National Science Foundation, *International Science and Technology Data Update: 1991*, p. 123; Patrick Oster, "U.S. Electronics Plug in to Western Europe," *Washington Post National Weekly Edition*, March 2–8, 1992, p. 22.

82. "Anchors Away, My Boys," *The Economist*, Jan. 12, 1991, p. 73; Sylvia Nasar, "American Revival in Manufacturing Seen in U.S. Report," *New York Times*, Feb. 5, 1991, p. 1.

83. Joseph S. Nye, Jr., *Bound to Lead: The Changing Nature of American Power* (New York: Basic Books, 1990), p. 72.

84. MIT Commission On Industrial Productivity, *Made in America: Regaining the Productive Edge* (Cambridge, Mass.: MIT Press, 1989); Clyde V. Prestowitz, Jr., *Trading Places: How We Allowed Japan to Take the Lead* (New York: Basic Books, 1988); Ezra F. Vogel, *Japan as Number One: Lessons for America* (Cambridge, Mass.: Harvard University Press, 1979); Daniel I. Okimoto, *Between MITI and the Market: Japanese Industrial Policy for High Technology* (Palo Alto, Calif.: Stanford University Press, 1989); Henry Ergas, "Does Technology Policy Matter?" in Guile and Brooks, *Technology and Global Industry;* EUREKA Secretariat, *EUREKA 1989 Project Progress Report* (Brussels, 1989).

85. Chalmers Johnson, ed., *The Industrial Policy Debate* (San Francisco: Institute for Contemporary Studies Press, 1984); F. Gerard Adams and Lawrence R.

Klein, eds., *Industrial Policies for Growth and Competitiveness* (Lexington, Mass.: Lexington Books, 1983); William Diebold, Jr., *Industrial Policy as an International Issue* (New York: McGraw-Hill, 1980); Competitiveness Policy Council, *Building a Competitive America: First Annual Report to the President and Congress* (Washington, D.C.: Competitive Policy Council, Mar. 1, 1992); Porter, *Competitive Advantage of Nations;* Kenichi Ohmae, *The Borderless World: Power and Strategy in the Interlinked Economy* (New York: Harper Business Press, 1990); George C. Lodge, *Perestroika for America: Restructuring U.S. Business-Government Relations for Competitiveness in the World Economy* (Boston: Harvard Business School Press, 1990).

86. Porter, *Competitive Advantage of Nations*, pp. 617–82.

87. Competitiveness Policy Council, *Building a Competitive America*, pp. 27–37.

88. Paul R. Krugman, "Introduction: New Thinking about Trade Policy," in Krugman, *Strategic Trade Policy*, p. 15.

89. It is noteworthy that, by one means or other, Japan has been able to maintain a 94 percent share of its own market in high-technology products, while the U.S. share of its own market has fallen since 1970 from 95 percent to 84 percent (National Science Board, *Science and Engineering Indicators, 1989*, app., table 7-5).

90. See the discussion of this difference of approach in chapter 2.

91. Wendy Schact, *Cooperative R&D: Federal Efforts to Promote Industrial Competitiveness*, CRS Issue Brief, Congressional Research Service, Library of Congress (Washington, D.C., Sept. 24, 1990); Office of Technology Assessment, *Competing Economies: America, Europe, and the Pacific Rim* (Washington, D.C.: OTA, Oct. 1991); Executive Office of the President, Office of Science and Technology Policy (OSTP), *U.S. Technology Policy*, Sept. 26, 1990. Lewis M. Branscomb discusses the OSTP policy statement, with views of his own, in an excellent summary piece, "Toward a U.S. Technology Policy," *Issues in Science and Technology* 7, no. 4 (Summer 1991): 50–55.

92. Thomas, C. Palmer, Jr., "Industrial Policy Inevitable for the US," *Boston Globe*, Mar. 8, 1992, pp. 73–74; Schact, *Cooperative R&D*, p. 5.

93. Office of Technology Assessment, *Nation's Stake in HDTV Is More than Just Another Pretty TV Picture* (Washington, D.C.: OTA, June 12, 1990); "The World at War," *The Economist*, Aug. 4, 1990, p. 58; Robert Buderi, "US Plan Just a Memory," *Nature* 343 (Jan. 25, 1990): 296.

94. Tyson, *Who's Bashing Whom?* chap. 2.

95. David C. Mowery and Nathan Rosenberg, *Technology and the Pursuit of Economic Growth* (New York: Cambridge University Press, 1989), p. 277; James Sterngold, "Intractable Trade Issues with Japan," *New York Times*, Dec. 4, 1991, p. D1.

96. In recognition of the new significance of domestic structures and patterns of behavior, the United States and Japan began negotiations in 1989 to attempt to agree on what elements were impeding free trade and how they might be changed. The negotiations, called the Structural Impediments Initiative (SII), have so far served only to highlight the depth of the problem (Steven R. Weisman, "Trade Talks Aim at Economic Habits," *New York Times*, Sept. 4, 1989; James Sterngold,

"U.S. and Japan Give Out Economic 'Report Cards,'" *New York Times*, May 23, 1991, p. D2).

97. Tyson, *Who's Bashing Whom?* chap. 2; C. Fred Bergston and Edward M. Graham, *Globalization of Industry and National Governments* (Washington, D.C.: Institute for International Economics, forthcoming, 1992).

98. An MIT faculty study analyzed the many issues involved in foreign access to American universities ("The International Relationships of MIT in a Technologically Competitive World," Massachusetts Institute of Technology, Cambridge, Mass., May 1, 1991).

99. The 1991 Science Indicators report of the U.S. National Science Board gives some disturbing data that shows such policy changes. ("U.S. R&D Expenditures in a Global Context," in *Science and Engineering Indicators, 1991*, NSB 91-1 [Washington, D.C., 1991], pp. 3–5).

100. A panel of the Office of Science and Technology Policy (OSTP) in the White House concludes there are twenty-two technologies that can be considered "critical." Most of those can more appropriately be thought of as "strategic" technologies—ones that are important for military and economic reasons but that do not have quite the unusual characteristics referred to here of being essential in themselves, necessary for the design of downstream products, and extremely difficult to recover once lost (OSTP, *Report of the National Critical Technologies Panel* [Washington, D.C., Mar. 1991]).

101. Michael Borrus, Laura Tyson, and John Zysman, "Creating Advantage: How Government Policies Shape International Trade in the Semiconductor Industry," in Krugman, *Strategic Trade Policy*, pp. 91–113.

102. Ibid., p. 93.

103. Quoted in David E. Sanger, "Building Smaller, Buying Bigger," *New York Times Magazine*, Feb. 18, 1990, p. 68.

104. Quoted in David E. Sanger, "In Computer Research Race, Japanese Increase Their Lead," *New York Times*, Feb. 21, 1990, p. D4.

105. Borrus, Tyson, and Zysman, "Creating Advantage," pp. 93–94.

106. There have been repeated charges that the Japanese have, in fact, denied or delayed chip deliveries and have made new models available to domestic manufacturers well before allowing them to be shipped to U.S. manufacturers (Michael Borrus, "Chips of State," *Issues in Science and Technology* 7, no. 1 [Fall 1990]: 43–44; Charles H. Ferguson, "Computers and the Coming of the U.S. Keiretsu," *Harvard Business Review*, July/Aug., 1990, pp. 64, 66).

107. The U.S. government's concern about this growing, and perhaps permanent, dependence on Japan for semiconductors essential to modern weaponry was reflected in two major studies in 1986, one by the Pentagon's Defense Science Board and the other in the National Security Council (Eliot Marshall, "Imported Chips: A Security Risk?" *Science* 232 [Apr. 4, 1986]: 12–13).

108. In 1989, the National Advisory Committee on Semiconductors, established by legislation, reflected a deep national concern over the issue in a report to the president and Congress (*A Strategic Industry at Risk* [Washington, D.C., Nov. 1, 1989]). The committee prepared to disband in 1992 out of frustration that none of its recommendations had been carried out (Andrew Pollack, "Frustrated, Chip Group May Disband," *New York Times*, Feb. 12, 1992, p. D5).

109. Borrus, "Chips of State," p. 40; Lawrence Edelman, "Policy Paralysis," *Boston Globe*, Dec. 3, 1989, p. A37.

110. "The Final Thrust," *The Economist*, Mar. 18, 1989, p. 116.

111. Andrew Pollack, "U.S. Chips Gain Is Japan's Loss," *New York Times*, Jan. 3, 1991, p. D1.

112. Andrew Pollack, "Slump Hits Japanese Electronics," *New York Times*, Feb. 24, 1992, p. D1.

113. "The Costly Race Chipmakers Can't Afford to Lose," *Business Week*, Dec. 10, 1990, p. 185.

114. "Final Thrust."

115. Graham and Krugman, *Foreign Direct Investment*, p. 70. The agreement was renewed in 1991 for a further three years (Keith Bradsher, "Chip Pact Set by U.S. and Japan," *New York Times*, June 4, 1991, p. D1).

116. Otis Port, "Sematech May Give America's Middleweights a Fighting Chance," *Business Week*, Dec. 10, 1990, p. 186.

117. David E. Sanger, "Contrasts on Chips," *New York Times*, Jan. 18, 1990, p. D1. Some suspect that the Japanese chip manufacturers boosted production in order to lower the price, and agreed to cut production and raise prices only after the attempt to create U.S. Memories failed (Borrus, "Chips of State"). Others strongly disagree with that interpretation (Kenneth Flamm in "Forum," *Issues in Science and Technology* 7, no. 2 [Winter 1990/91]: 10). It is indicative of IBM's view of the matter that it would have been willing, in an unprecedented move, to make its own advanced technology available to U.S. Memories if that venture had come into being (Ferguson, "Computers," p. 67).

118. "JESSI Enters Crucial Phase," *Nature* 341 (Oct. 19, 1989): 559; Borrus, "Chips of State," p. 43.

119. National Advisory Committee on Semiconductors, *A Strategic Industry at Risk*, p. 1.

120. Sanger, "Japanese Increase Their Lead." IBM is the sole U.S. entry in this competitive race and is gambling large sums on the future importance of this technology (John Markoff, "IBM's Big Gamble on X-Rays," *New York Times*, July 19, 1991, p. D1).

121. Sanger, "Japanese Increase Their Lead" Port, "Sematech," p. 186.

122. Port, "Sematech," p. 186; John Markoff, "A.T.&T. Making Comeback in Chips," *New York Times*, June 25, 1990, p. D1.

123. Charles Ferguson asserts this argument is wrong, contending that the United States will not be able to sustain the necessary r/d if DRAM manufacturing capacity falls and if the Japanese prove to be as adept at catching up in microprocessors as they did in DRAMs ("Computers," p. 66).

124. Andrew Pollack, "Gridlock at the Chip-to-Microprocessor Intersection," *New York Times*, Mar. 22, 1992, p. 11.

125. John Markoff, "IBM Joins Siemens in Developing Chips," *New York Times*, Jan. 25, 1990, p. D1; idem, "Motorola Will Make Chips with Toshiba at Japanese Site," *New York Times*, Dec. 14, 1990, p. D1.

126. Analysts appear divided on this point. Yves Doz argues that r/d tends to remain centralized in the home country, notwithstanding growing incentives for decentralization ("International Industries," p. 106). Graham and Krugman assert,

on the other hand, that "the headquarters effect is not strong. . . . multinationals . . . show little tendency to concentrate R&D at home" (*Foreign Direct Investment*, p. 53).

127. Borrus, "Chips of State," pp. 43–44.

128. Harlan Cleveland, "Information as a Resource," *The Futurist*, Dec. 1982; Anne Wells Branscomb, "Property Rights in Information," in Guile, *Information Technologies*; Office of Technology Assessment, *Computer Software and Intellectual Property*, Background Paper (Washington, D.C.: OTA, 1990).

129. John Burgess, "Making Sure the World Pays for American Ideas," *Washington Post National Weekly Edition*, Dec. 21, 1987, p. 20.

130. United Nations Conference on Trade and Development (UNCTAD), Secretariat, "Strategy for the Technological Transformation of Developing Countries," in Pradip K. Ghosh, ed., *Technology Policy and Development: A Third World Perspective* (Westport, Conn.: Greenwood Press, 1984), pp. 218–28; Overseas Development Council, *Intellectual Property and Developing Countries: Options for U.S. Policy*, ODC Policy Focus No. 5 (Washington, D.C.: ODC, 1989).

131. John A. Armstrong, "Trends in Global Science and Technology and What They Mean for Intellectual Property Systems," paper prepared for National Research Council conference, "Global Dimensions of Intellectual Property Rights in Science and Technology," Washington, D.C., Jan. 8–9, 1992.

132. Abel, *The Shattered Bloc*; Robert Heilbroner, "Reflections: After Communism," *New Yorker*, Sept. 10, 1990, pp. 91–100.

133. There are many analyses of the science- and technology-related problems in the Soviet and Eastern European economies. See particularly Graham, *Science and the Soviet Social Order*; Ronald Amann and Julian Cooper, eds., *Technological Progress and Soviet Economic Development* (Oxford: Blackwell, 1986); Zhores Aleksandrovich Medvedev, *Soviet Science* (New York: Norton, 1978); and Joseph S. Berliner, *The Innovation Decision in Soviet Industry* (Cambridge, Mass.: MIT Press, 1976).

134. "Command economy" is used here to signify a centrally planned and centrally controlled economy. There is obviously a broad spectrum running from market to command economies, but our purpose is to emphasize general principles rather than the variance to be found in reality.

135. Nathan Rosenberg and L. E. Birdzell, Jr., *How the West Grew Rich: The Economic Transformation of the Industrial World* (New York: Basic Books, 1986), p. 244.

136. Harry Harding and Ed A. Hewett, "Socialist Reforms and the World Economy," in John D. Steinbruner, ed., *Restructuring American Foreign Policy* (Washington, D.C.: Brookings Institution, 1989), p. 163.

137. S. E. Goodman and W. K. McHenry, "Computing in the USSR: Recent Progress and Policies," *Soviet Economy* 2, no. 4 (1986): 327–54.

138. Marshall D. Shulman, "The Superpowers: Dance of the Dinosaurs," *Foreign Affairs* 66, no. 3 (*America and the World, 1987/88*): 494–515.

139. Goodman, "Information Technologies and the Citizen," pp. 51–67.

140. Medvedev, *Soviet Science*; Graham, *Science and the Soviet Social Order*; Michael Kaser, "The Impact of Technological Change on East-West Economic Relations," in F. Stephen Larrabee, ed., *Technology and Change in East-West*

Relations, East-West Monograph Series, no. 6 (New York: Institute for East-West Security Studies, 1988), pp. 147–63.

141. Mikhail Gorbachev, *Perestroika: New Thinking for Our Country and the World* (New York: Harper & Row, 1987). *Perestroika* means, roughly, restructuring.

142. Loren Graham, "Gorbachev's Great Experiment," *Issues in Science and Technology* 4, no. 2 (Winter 1988): 23–32.

143. Shulman, "Superpowers," p. 497.

144. Robert G. Kaiser, "Gorbachev: Triumph and Failure," *Foreign Affairs* 70, no. 2 (Spring 1991): 160–74.

145. Mandelbaum, "Coup de Grace"; Dimitri Simes, "Russia Reborn,' *Foreign Policy*, no. 85 (Winter 1991/92): 41–62.

146. Mandelbaum, "Coup de Grace"; Gail W. Lapidus, "Gorbachev's Nationalities Problem," *Foreign Affairs* 68, no. 4 (Fall 1989): 92–108; Valerii Tishkov, "*Glasnost* and the Nationalities within the Soviet Union," *Third World Quarterly* 11, no. 4 (Oct. 1989): 191–207; Zbigniew Brzezinski, "Post-Communist Nationalism," *Foreign Affairs* 68, no. 5 (Winter 1989/90): 1–25.

147. Mandelbaum, "Coup de Grace," p. 181.

148. Daniel Bell, *The End of Ideology: On the Exhaustion of Political Ideas in the Fifties*, with a new afterword (Cambridge, Mass.: Harvard University Press, 1988); Francis Fukuyama, "The End of History," *National Interest*, Summer 1989, pp. 3–35.

149. There is a massive literature dealing with development; the part that is closely related to technology is generally organized by specific subjects, such as food or energy. General works that attempt to explore the overall role of technology as a whole are quite thin in number. See especially Daniel Lerner, *The Passing of Traditional Society: Modernizing the Middle East* (Glencoe, Ill.: Free Press, 1958); Graham Jones, *The Role of Science and Technology in Developing Countries* (Oxford: Oxford University Press, 1971); Jairam Ramesh and Charles Weiss, Jr., eds., *Mobilizing Technology for World Development* (New York: Praeger, 1979); Robert P. Morgan, *Science and Technology for International Development: An Assessment of U.S. Policies and Programs* (Boulder, Colo.: Westview Press, 1984); Ghosh, *Technology Policy and Development*; Francisco R. Sagasti, *Technology, Planning, and Self-Reliant Development: A Latin American View* (New York: Praeger, 1979); and Martin Fransman, *Technology and Economic Development* (Boulder, Colo.: Westview Press, 1986).

150. Kuznets, *Modern Economic Growth*, pp. 9–10.

151. The continued "brain drain" from developing countries, in particular to the United States, is a constantly festering issue. It is made more difficult because the United States has come to depend on foreign-born scientists and engineers to staff its industry and universities (National Academy of Engineering, *Foreign and Foreign-Born Engineers in the U.S.: Infusing Talent, Raising Issues* [Washington, D.C.: National Academy Press, 1988]; National Science Board, *Report of the NSB Committee on Foreign Involvement in U.S. Universities*, NSB 89–80 [Washington, D.C., 1989]).

152. Eugene B. Skolnikoff, "Transfer of Technology: The Economic and Security Debate," paper presented before the Chicago Council on Foreign Relations,

Chicago, Ill., May 13, 1982; Charles Weiss, Jr., "Scientific and Technological Constraints to Economic Growth and Equity," in Robert E. Evenson and Gustav Ranis, eds., *Science and Technology: Lessons for Development* (Boulder, Colo.: Westview Press, 1990), pp. 17–41.

153. Skolnikoff, "Transfer of Technology," p. 14. In explaining the simultaneous appearance of inventions in different parts of the world, historian Arnold Pacey notes, "But quite often, the most important factor was that the achievements of one society stimulated people elsewhere to make different but related inventions" (*Technology in World Civilization*, p. vii).

154. Merton J. Peck and Akira Goto, "Technology and Economic Growth: The Case of Japan," *Research Policy* 10 (1981): 222–43; Frances Stewart, "Arguments for the Generation of Technology by Less-Developed Countries," *Annals* (American Academy of Political and Social Sciences) 458 (Nov. 1981): 97; David J. Teece, "The Market for Know-how and the Efficient International Transfer of Technology," *Annals* 458 (Nov. 1981): 81–96; Nathan Rosenberg, *Perspectives on Technology* (New York: Cambridge University Press, 1976), pp. 151–72.

155. Charles Weiss offers some detailed guidance for allocation of resources for developing countries, with an approach based on determining "stages" of scientific and technological development ("Scientific and Technological Constraints"). See also M. R. Bhagavan, *The SAREC Model: Institutional Cooperation and the Strengthening of National Research Capacity in Developing Countries*, 1992:1 (Stockholm: Swedish Agency for Research Cooperation with Developing Countries, 1992).

156. U.S. technical-assistance activities in fact began in the late 1930s with the Good Neighbor policy toward Latin America, a program headed by Nelson Rockefeller (Robert Clarke Stowe, "Agricultural Politics and Technical Assistance for Development," [Ph.D. diss., Massachusetts Institute of Technology, Department of Political Science, Feb. 1990]). It is not often appreciated that the United States was actually the beneficiary in the nineteenth century of technical assistance from Europe.

157. Morgan, *Science and Technology for International Development*.

158. Eugene B. Skolnikoff, *Science, Technology, and American Foreign Policy* (Cambridge, Mass.: MIT Press, 1967), pp. 152–59. Most agreements in science and technology involving the United States are given in the so-called Title V reports (U.S. House of Representatives, Committee on Foreign Affairs and Committee on Science and Technology, *Annual Reports Submitted to the Congress by the President, Pursuant to Section 503(b) of Title V of PL 95–426*, Joint Committee Prints). At times, agreements for scientific or technological cooperation were reached to avoid the embarrassment of a meeting of senior leaders that had no other positive outcomes.

159. Dependency theorists argue that North-South dependencies have a more basic cause than interdependence—namely, that third-world economies are in effect the vassals of capitalist economies, serving as the suppliers of low-cost labor or raw materials and enmeshed in a system that denies them the opportunity for autonomous development or advancement. See Theotonio Dos Santos, "The Structure of Dependence," *American Economic Review* 60, no. 2 (1965); Cardoso and Faletto, *Dependency and Development in Latin America*; and Peter Evans,

Dependent Development: The Alliance of Multinational, State, and Local Capital (Princeton: Princeton University Press, 1979). Whether the particular kinds of dependencies between countries of North and South are a result of exploitation or are due to more ideologically neutral causes is not immediately relevant to our inquiry.

160. Antonio José J. Botelho, "Brazil's Independent Computer Strategy: Brazil Has Developed Its Computer Industry by Restricting Imports," *Technology Review* 90, no. 4 (May/June 1987): 36–45; Paulo Bastos Tigre, *Technology and Competition in the Brazilian Computer Industry* (New York: St. Martin's Press, 1983); Ricardo Bonalume, "Opening Up to Computers," *Nature* 352 (Aug. 15, 1991): 558.

161. Natural resources are discussed in greater detail later in this chapter.

162. Marcellus S. Snow, *The International Telecommunications Satellite Organization (Intelsat): Economic and Institutional Challenges Facing an International Organization* (Baden-Baden: Nomos Verlagsgesellschaft, 1987); Wilson P. Dizard, Jr., "International Policy Issues in Satellite Communications," *Journal of International Affairs* 39 (Summer 1985): 121–28.

163. National Research Council, Space Applications Board, *Remote Sensing of the Earth from Space: A Program in Crisis* (Washington, D.C.: National Academy Press, 1985); NASA Advisory Council, Task Force on International Relations in Space, *International Space Policy for the 1990s and Beyond* (Washington, D.C., Oct. 12, 1987); Maria Carmen de Mello Lemos, "Landsat as a Commercial Enterprise" (Master's thesis, Massachusetts Institute of Technology, Department of Political Science, June 1990); "New Lease for Landsat," *Science* 254 (Dec. 13, 1991): 1585.

164. The United States has a decidedly mixed reputation for reliability as a partner in international cooperation. See Alexander Keynan, *The United States as a Partner in Scientific and Technological Cooperation: Some Perspectives from across the Atlantic* (New York: Carnegie Commission on Science, Technology, and Government, June 1991), pp. 41–74.

165. Tom Logsdon, *Space Inc.: Your Guide to Investing in Space Exploration* (New York: Crown Publishers, 1988); Michael Harr and Rajiv Kohli, *Commercial Utilization of Space: An International Comparison of Framework Conditions* (Columbus, Ohio: Battelle Press, 1990); G. Duchossois, "Remote Sensing from Space: European Achievements and Prospects," *Impact of Science on Society*, no. 140 (1985): 319–35.

166. Representative examples of such formulations can be found in Bruce Russett and Harvey Starr, *World Politics: The Menu for Choice* (New York: W. H. Freeman, 1985); John Spanier, *Games Nations Play: Analyzing International Politics*, 4th ed. (New York: Holt, Rinehart & Winston, 1981); Daniel S. Papp, *Contemporary International Relations: Frameworks for Understanding*, 2d ed. (New York: Macmillan, 1988); and James E. Dougherty and Robert L. Pfaltzgraff, *Contending Theories of International Relations: A Comprehensive Survey*, 3d ed. (New York: Harper & Row, 1990).

167. The need for tin for bronze weapons is postulated as a cause of the Trojan War in Derek Ager, "Ore That Launched a Thousand Ships," *New Scientist* 106 (June 20, 1985): 28. Daniel Yergin traces the role of oil in the origins of World War II in the Pacific in *The Epic Quest for Oil, Money, and Power* (New York: Simon

& Schuster, 1991). Recent analyses of the relation of natural resources to international affairs include Nazli Choucri and Robert C. North, *Nations in Conflict: National Growth and International Violence* (San Francisco: W. H. Freeman, 1975); Hanns W. Maull, *Energy, Minerals, and Western Security* (Baltimore: Johns Hopkins University Press, 1984); and Philip Connelly and Robert Perlman, *The Politics of Scarcity: Resource Conflicts in International Relations*, Royal Institute of International Affairs (Oxford: Oxford University Press, 1975).

168. Allen V. Kneese, *The Economics of Natural Resources*, Reprint no. 243 (Washington, D.C.: Resources for the Future, 1989); V. Kerry Smith and John V. Krutilla, eds., *Explorations in Natural Resource Economics* (Baltimore: Johns Hopkins University Press, 1982); Ferdinand E. Banks, *The Economics of Natural Resources* (New York: Plenum Press, 1976); Charles W. Howe, *Natural Resource Economics: Issues, Analysis, and Policy* (New York: John Wiley & Sons, 1979).

169. Rosenberg, *Perspectives on Technology*, p. 234.

170. Arden L. Bement, Jr., "Materials Sector Profile," in Anne G. Keatley, ed., *Technological Frontiers and Foreign Relations* (Washington, D.C.: National Academy Press, 1985); A. J. van Griethuysen, *New Applications of Materials* (The Hague: Scientific and Technical Press, 1988); Peter A. Psaras and H. Dale Langford, eds., *Advancing Materials Research*, National Academy of Engineering (Washington, D.C.: National Academy Press, 1987); John I. Brauman, "Frontiers in Materials Science," *Science* 255 (Feb. 28, 1992): 1049; *Frontiers in Materials Science*, special section in *Science* 255 (Feb. 28, 1992): 1077–1112.

171. Peter F. Drucker, "The Changed World Economy," *Foreign Affairs* 64, no. 4 (Spring 1986): 768; Gerd Junne et al., "Dematerialization of Production: Impact on Raw Material Exports of Developing Countries," *Third World Quarterly* 11, no. 2 (Apr. 1989): 128–42.

172. Dependencies are not necessarily undesirable; they are, in fact, inevitable. The geopolitical concern is with ones that create undesirable vulnerabilities or ones that have major international economic implications.

173. Anthony H. Cordesman, *The Gulf and the Search for Strategic Stability* (Boulder, Colo.: Westview Press, 1984), p. 729.

174. A particularly careful and balanced assessment of the consequences of rapid population growth is given in Geoffrey McNicoll, "Consequences of Rapid Population Growth: An Overview and Assessment," *Population and Development Review* 10, no. 2 (June 1984): 177–240. For more alarmist views, see Paul R. Ehrlich, *The Population Bomb* (New York: Ballantine, 1968); and Paul R. Ehrlich and Anne H. Ehrlich, *The Population Explosion* (New York: Simon and Schuster, 1990). A dissenting view to the alarmist perspective is expressed in Julian L. Simon, *The Ultimate Resource* (Princeton: Princeton University Press, 1981).

175. Carl Wahren, "Population and Development: The Burgeoning Billions," *OECD Observer*, Dec. 1988/Jan. 1989, p. 7.

176. Ibid., p. 5.

177. Adi Ignatius, "Beijing Boom: China's Birthrate Rises Again despite a Policy of One-Child Families," *Wall Street Journal*, July 14, 1988, p. 1.

178. UN Population Fund, *The State of World Population* (New York: UN Population Fund, 1991); Steven W. Sinding and Sheldon J. Segal, "Birth-Rate News," *New York Times*, Dec. 19, 1991, p. A31.

179. Science and technology are sometimes assumed to be the "cause" of the population explosion, through medical advances that cut mortality rates. In fact, the causes are much more complex; the steep population increases in the last two centuries were apparently more a product of the industrial revolution than of medical and sanitation advances (Kingsley Davis, "The History of Birth and Death," *Bulletin of the Atomic Scientists* 42, no. 4 [Apr. 1986]: 20–23; Paul R. Ehrlich, Anne H. Ehrlich, and John P. Holdren, eds., *Ecoscience: Population, Resources, and Environment*, 3d ed., rev. [San Francisco: W. H. Freeman, 1977], pp. 189–202).

180. R. J. Lapham and G. B. Simmons, eds., *Organizing for Effective Family Planning Programs* (Washington, D.C.: National Academy Press, 1987); Robert Repetto, "Population, Resource Pressures, and Poverty," in Robert Repetto, ed., *The Global Possible: Resources, Development, and the New Century* (New Haven, Conn.: Yale University Press, 1985), pp. 131–69; Sinding and Segal, "Birth-Rate News."

181. Rosenau, *Turbulence in World Politics*, p. 88.

182. Harold Sprout, "Geopolitical Hypotheses in Technological Perspective," *World Politics* 15, no. 2 (Jan. 1963): 187–212.

183. National Commission on Excellence in Education, *A Nation at Risk: The Imperative for Educational Reform: A Report to the Nation and the Secretary of Education* (Washington, D.C.: The Commission, 1983); Senta A. Raizen and Lyle V. Jones, eds., *Indicators of Precollege Education in Science and Mathematics: A Preliminary Review*, National Research Council, Committee on Indicators of Precollege Science and Mathematics Education (Washington, D.C.: National Academy Press, 1985); Office of Technology Assessment, *Educating Scientists and Engineers: Grade School to Grad School* (Washington, D.C.: OTA, 1988).

184. Maria Maguire, "Making Provision for Ageing Populations," *OECD Observer*, Oct./Nov. 1987, pp. 4–9. The fall in the birthrate in Japan has caused such consternation that the Japanese cabinet discussed whether too many women were receiving higher education, thus reducing their time for, or interest in, childbearing (David E. Sanger, "Minister Denies He Opposed College for Women," *New York Times*, June 19, 1990, p. A2). East Asian policy responses are analyzed in Linda G. Martin, "Population Aging Policies in East Asia and the United States," *Science* 251 (Feb. 1, 1991): 527–31.

185. Jane Mencken, ed., *World Population and U.S. Policy: The Choices Ahead*, Report of the American Assembly, Columbia University (New York: W. W. Norton, 1986).

186. The 1990 change in U.S. immigration law was expressly intended to make it easier to recruit people with essential skills from abroad (Seth Mydans, "For Skilled Foreigners, Lower Hurdles to the U.S.," *New York Times*, Nov. 5, 1990, p. A12).

187. "Race, Votes, and Power: The Numbers Game," *The Economist*, Dec. 26, 1987, pp. 63–66. The fighting between the Muslims in Azerbaijan and the Christians in Armenia and among ethnic populations in other Soviet republics is almost certainly a harbinger of future conflict (Hamish McDonald, "Tremors in Tartary: Islamic Republics Assert Identity as Moscow Falters," *Far Eastern Economic Review* 148 [May 24, 1990]: 27–28).

188. Lester R. Brown, "The Growing Grain Gap," *World-Watch* 1, no. 5 (Sept./ Oct. 1988): 16; World Commission on Environment and Development, *Our Common Future* Oxford: Oxford University Press, 1987), p. 118.

189. A. Hunter Dupree, *Science in the Federal Government: A History of Policies and Activities to 1940* (Cambridge, Mass.: Harvard University Press, Belknap Press, 1957), pp. 149–83.

190. Lester R. Brown, *The Green Revolution and Development in the 1970s* (New York: Praeger, 1970); Thomas T. Poleman and Donald K. Freebairn, eds., *Food, Population, and Employment: The Impact of the Green Revolution* (New York: Praeger, 1973); Andrew C. Pearse, *Seeds of Plenty, Seeds of Want: Social and Economic Implications of the Green Revolution* (UN Research Institute for Social Development, Oxford: Oxford University Press, Clarendon Press, 1980).

191. World Commission on Environment and Development, *Our Common Future*, pp. 100, 118.

192. Brown, "The Growing Grain Gap," p. 15.

193. World Commission on Environment and Development, *Our Common Future*, p. 120.

194. Pearse, *Seeds of Plenty*.

195. Warren C. Baum with the collaboration of Michael L. LeJeune, *Partners against Hunger: The Consultative Group on International Agricultural Research* (Washington, D.C.: World Bank, 1986).

196. Brown, *Green Revolution*, pp. 77–100; Poleman and Freebairn, *Food, Population, and Employment*, pp. 97–119; Pearse, *Seeds of Plenty*, pp. 173–82.

197. Lester R. Brown, Christopher Flavin, and Sandra Postel, "A World at Risk," in Worldwatch Institute, *State of the World, 1989: Report on Progress toward a Sustainable Society* (New York: W. W. Norton, 1989), p. 12.

198. Brown, "The Growing Grain Gap," p. 11.

199. Paul Lewis, "Peril of Third-World Famine Is Seen by U.N. Food Agency," *New York Times*, Mar. 27, 1990, p. A10; U.S. Department of Agriculture, Foreign Agriculture Service, *World Grain Situation and Outlook* (Washington, D.C., 1989).

200. Brown, "The Growing Grain Gap," pp. 11–12.

201. Brown, Flavin, and Postel, "A World at Risk," p. 15.

202. U.S. Department of Agriculture, Economic Research Service, *Global Food Assessment, Situation and Outlook Report*, GFA 1, Nov. 1990, p. 9.

203. Michael Lipton with Richard Longhurst, *New Seeds and Poor People* (Baltimore: Johns Hopkins University Press, 1989), pp. 368, 370–71; Jack Ralph Kloppenburg, *First the Seed: The Political Economy of Plant Biotechnology, 1492–2000* (New York: Cambridge University Press, 1988); Lawrence Busch et al. *Plants, Power, and Profit: Social, Economic, and Ethical Consequences of the New Biotechnologies* (Cambridge: Blackwell, 1991); Carliene Brenner, "Biotechnology in the Developing World: Lessons from Maize," *OECD Observer*, Aug./Sept. 1991, pp. 9–12; Anne S. Moffat, "Biotechnology Reaches beyond the High-Tech West," *Science* 225 (Feb. 21, 1992): 919.

204. Kloppenburg, *First the Seed*, pp. 162–63; Sir Otto H. Frankel, "Genetic Dangers in the Green Revolution," *World Agriculture* 19, no. 3 (July 1970): 9–13;

Cary Fowler and Pat Mooney, *Shattering: Food, Politics, and the Loss of Genetic Diversity* (Tucson, Ariz.: University of Arizona Press, 1990).

205. Fowler and Mooney, *Shattering*, pp. 42–45; Harry Serlis, *Wine in America* (New York: Newcomen Society, 1972), p. 10.

206. Rye fungus that caused temporary madness may have developed under unusual weather conditions in France in 1789. It has been cited as the probable cause of widespread terror in the countryside, a pivotal event in the history of the French Revolution (Mary Kilbourne Matossian, *Poisons of the Past: Molds, Epidemics, and History* [New Haven, Conn.: Yale University Press, 1989], pp. 81–87).

207. Steve Raynor, "The Greenhouse Effect in the US: The Legacy of Energy Abundance," in Michael Grubb et al., *Country Studies and Technical Options*, vol. 2 of *Energy Policy and the Greenhouse Effect*, Royal Institute of International Affairs (Aldershot, England: Dartmouth Publishing, 1991), pp. 250–51.

208. Frank Close, *Too Hot to Handle: The Race for Cold Fusion* (Princeton: Princeton University Press, 1991).

209. H. G. Wells, *The World Set Free* (New York: E. P. Dutton, 1914); Harold Nicolson, *Public Faces* (Boston: Houghton Mifflin, 1932). Enrico Fermi had seen the evidence of the splitting of the atom several years earlier, but that evidence ran counter to the existing paradigm to the extent that neither he nor others correctly interpreted their observations. Much earlier, the distinguished atomic scientist Ernest Rutherford had predicted that man would never be able to tap the energy of the atom (Richard Rhodes, *The Making of the Atomic Bomb* [New York: Simon & Schuster, 1986], pp. 38, 208).

210. Rhodes, *Atomic Bomb*, pp. 428–42.

211. Richard Saltus, "Pa. Plant's Experience May Be Model for Yankee Decommissioning," *Boston Globe*, Feb. 27, 1992, p. 68.

212. Robert C. Williams and Philip L. Cantelon, eds., *The American Atom: A Documentary History of Nuclear Policies from the Discovery of Fission to the Present, 1939–1984* (Philadelphia: University of Pennsylvania Press, 1985), pp. 294–95.

213. Attributed to Lewis Strauss, member and sometime chairman of the Atomic Energy Commission, as noted in Daniel Ford, *The Cult of the Atom: The Secret Papers of the Atomic Energy Commission* (New York: Simon & Schuster, 1982), p. 14.

214. Richard G. Hewlett and Oscar E. Anderson, Jr., *The New World, 1939/ 1946*, vol. 1 of *A History of the United States Atomic Energy Commission* (University Park, Pa.: Pennsylvania State University Press, 1962), p. 327. In a bizarre sequel, the record of that private 1943 agreement was "lost" for many years when it was misfiled in the White House (ibid., p. 428).

215. Andrew J. Pierre, *Nuclear Politics: The British Experience with an Independent Strategic Force, 1939–1970* (London: Oxford University Press, 1972), pp. 136–39.

216. James R. Schlesinger, "Atoms for Peace Revisited," in Joseph F. Pilat, Robert E. Pendley, and Charles K. Ebinger, eds., *Atoms for Peace: An Analysis after Thirty Years* (Boulder, Colo.: Westview Press, 1985), pp. 5–15; Richard G. Hewlett, "From Proposal to Program," in Pilat, Pendley, and Ebinger, *Atoms for Peace*, pp. 25–33.

217. Yergin, *The Prize*, 306–50.

218. Jacques de la Ferté, "What Future for Nuclear Power?" *OECD Observer*, Apr./May 1990, pp. 26–30; "Gas Bubbles from the Mutsu," *The Economist*, Apr. 14, 1990, p. 34.

219. Terence Price, "Politics of Electricity Production," *Nature* 351 (June 6, 1991): 435–36.

220. For an excellent summary of the history of the nuclear-power industry and its difficulties, see Irvin C. Bupp and Jean-Claude Derian, *Light Water: How the Nuclear Dream Dissolved* (New York: Basic Books, 1978).

221. Ernest J. Sternglass, *Low-Level Radiation* (New York: Ballantine, 1972); Leslie Roberts, "British Radiation Study Throws Experts into Tizzy," *Science* 248 (Apr. 6, 1990): 24–25.

222. Dorothy Nelkin, *The Atom Besieged: Extraparliamentary Dissent in France and Germany* (Cambridge, Mass.: MIT Press, 1981); Michael R. Reich, "Mobilizing for Environmental Policy in Italy and Japan," *Comparative Politics* 16, no. 4 (July 1984): 379–402. A serious nuclear accident that took place in Japan in 1991 could lead to more opposition to nuclear power (David E. Sanger, "Japan Nuclear Accident May Impede Push for Plants," *New York Times*, Feb. 11, 1991, p. A3).

223. Kent Hansen et al., "Making Nuclear Power Work: Lessons from around the World," *Technology Review* 92, no. 2 (Feb./Mar., 1989): 30.

224. Thomas Land, "East Europe's Reactors in Trouble," *Nature* 355 (Jan. 9, 1992): 98; "Nuclear Perils in Eastern Europe," *The Economist*, July 27, 1991, p. 48; Grigorii Medvedev, *The Truth about Chernobyl*, trans. Evelyn Rossiter (New York: Basic Books, 1991); Zhores A. Medvedev, *The Legacy of Chernobyl* (New York: W. W. Norton, 1990); Iurii Shcherbak, *Chernobyl: A Documentary Story*, trans. Ian Press (New York: St. Martin's Press, 1989); O. M. Kovalevich, V. A. Sidorenko, and N. A. Shteinberg, "Soviet Bureaucracy and Nuclear Safety," *Forum for Applied Research and Public Policy* (University of Tennessee), Summer 1991, pp. 88–92.

225. Jon M. Van Dyke, "Ocean Disposal of Nuclear Wastes," *Marine Policy* 12, (Apr. 1988): 82–95.

226. Luther J. Carter, *Nuclear Imperatives and Public Trust: Dealing with Radioactive Waste* (Washington, D.C.: Resources for the Future, 1987); E. William Colglazier, Jr., ed., *The Politics of Nuclear Waste*, Aspen Institute (New York: Pergamon Press, 1982); Taylor Moore, "The Hard Road to Nuclear Waste Disposal," *EPRI Journal* (Electric Power Research Institute) 15, no. 4 (July/Aug. 1990): 4–18; Gerald Jacob, *Site Unseen: The Politics of Siting A Nuclear Waste Repository* (Pittsburgh, Pa.: University of Pittsburgh Press, 1990); Andrew Blowers, David Lowry, and Barry D. Solomon, *The International Politics of Nuclear Waste* (New York: St. Martin's Press, 1991).

227. "Closing the fuel cycle" refers to separation of the newly created plutonium after enriched uranium is burned in an ordinary reactor, so as to use the plutonium directly as fuel or in a breeder reactor—which in turn can produce additional fuel. This practice would serve as a way of multiplying the energy potential of natural uranium resources. Closing the cycle was at first assumed to be necessary to make nuclear power economic, but was later seen as unnecessary, at

least on economic grounds (Nuclear Energy Policy Study Group, *Nuclear Power Issues and Choices* [Cambridge, Mass.: Ballinger, 1977], pp. 65–66).

228. Hewlett and Anderson, *The New World*, pp. 531–619; Lawrence Scheinman, *The International Atomic Energy Agency and World Nuclear Order*, Resources for the Future (Baltimore: Johns Hopkins University Press, 1987), pp. 49–56.

229. Eugene B. Skolnikoff, *The International Imperatives of Technology: Technological Development and the International Political System*, Research Series, no. 16, Institute of International Studies, University of California, Berkeley, 1972, pp. 118–19.

230. Joseph A. Yager, *International Cooperation in Nuclear Energy* (Washington, D.C.: Brookings Institution, 1981), pp. 25–40; Scheinman, *International Atomic Energy Agency*; Lawrence Scheinman, *The Nonproliferation Role of the International Atomic Energy Agency*, Resources for the Future (Baltimore: Johns Hopkins University Press, 1985).

231. Scheinman, *Nonproliferation Role*.

232. Williams and Cantelon, *American Atom*, pp. 104–11.

233. Leonard Weiss, "Atoms for Peace and Nuclear Proliferation," in Pilat, Pendley, and Ebinger, *Atoms for Peace*, pp. 131–41; Warren H. Donnelly, "Agency Safeguards: A Model for Arms Control Verification," in Pilat, Pendley, and Ebinger, *Atoms for Peace*, pp. 251–60.

234. Lawrence M. Lidsky, "Nuclear Power: Levels of Safety," *Radiation Research* 113 (1988): 217–26.

235. Ford, *Cult of the Atom*.

236. Walter A. MacDougall, " . . . *The Heavens and the Earth: A Political History of the Space Age* (New York: Basic Books, 1985), pp. 276–97.

237. Ironically, the Soviets were able to launch larger satellites earlier than the United States *because* of their lag in technology. Their inferior nuclear-warhead capability required them to develop larger rockets to carry heavier warheads; the United States could design smaller missiles to carry equivalent firepower. As a result, the Soviets had larger missiles available for space launches than did the United States.

238. Skolnikoff, *Science, Technology, and American Foreign Policy*, pp. 223–48.

239. Eugene B. Skolnikoff, "Report of the Second Advisory Panel Meeting, Nov. 1983," in *Civilian Space Stations and the U.S. Future in Space* (Washington, D.C.: Office of Technology Assessment, Nov. 1984), pp. 133–39.

240. Alic et al., *Beyond Spinoff*, chapter 3; Roger H. Bezdek and Robert M. Wendling, "Sharing Out NASA's Spoils," *Nature* 355 (Jan. 9, 1992): 105–6.

241. Office of Technology Assessment, *Exploring the Moon and Mars: Choices for the Nation*, (Washington, D.C.: OTA, July 1991).

242. Hernán Garrido-Lecca Montáñez, "The Economics and Politics of Slot and Frequency Allocation in the Geostationary Orbit" (Master's thesis, Massachusetts Institute of Technology, Department of Political Science, Sept. 1990).

243. Office of Science and Technology Policy, Committee on Earth and Planetary Sciences, *Our Changing Planet: The FY 1993 U.S. Global Change Research Program*, Washington, D.C., 1992.

244. Office of Technology Assessment, *Exploring the Moon and Mars*, p. 1.

245. Warren E. Leary, "U.S. Advisers Urge Sweeping Change in Space Program," *New York Times*, Dec. 11, 1990, p. A1; John Noble Wilford, "Giving New Purpose to the Space Program," *New York Times*, Dec. 11, 1990, p. C14.

246. Jonathan N. Goodrich, *The Commercialization of Outer Space: Opportunities and Obstacles for American Business* (New York: Quorum Books, 1989), pp. 128–29.

247. Skolnikoff, "Second Advisory Panel," pp. 136–37; Spark M. Matsunaga, *The Mars Project: Journeys Beyond the Cold War* (New York: Hill & Wang, 1986). The history of cooperation between the United States and Europe on a manned space station has shown some of the perils of cooperation (James R. Asker, "Japanese and Europeans Irked by Latest Space Station Changes," *Aviation Week and Space Technology*, Nov. 6, 1989, pp. 19–22; Helen Garaghan, "Downgraded Space Station Dismays U.S.'s Partners," *New Scientist* Aug. 26, 1989, p. 20; "The Cost of Freedom," *The Economist*, Oct. 7, 1989, pp. 105–6; Keynan, *The United States as a Partner*, pp. 47–53).

248. John M. Logsdon, one of the more perceptive and influential analysts if U.S. space policy, generally agrees with this assessment, but argues that more-fundamental political issues are likely to emerge around space activities in the future ("Outer Space and International Space Policy: The Rapidly Changing Issues," in Daniel S. Papp and John R. McIntyre, eds., *International Space Policy: Legal, Economic, and Strategic Options for the Twentieth Century and Beyond* [New York: Quorum Books, 1987], pp. 31–41).

249. Murray Feshbach and Alfred Friendly, *Ecocide in the USSR* (New York: Basic Books, 1992); Fred Singleton, "Eastern Europe: Do the Greens Threaten the Reds?" *The World Today* 42 (Aug./Sept. 1986): 159–62; Richard Crampton, "The Intelligentsia, the Ecology, and the Opposition in Bulgaria," *The World Today* 46 (Feb. 1990): 23–26; Robert G. Darst, Jr., "Environmentalism in the USSR: The Opposition to the River Diversion Projects," *Soviet Economy* 4, no. 3 (1988): 223–52; Hilary F. French, *Green Revolutions: Environmental Reconstruction in Eastern Europe and the Soviet Union*, Worldwatch Paper 99 (Washington, D.C.: Worldwatch Institute, Nov. 1990).

250. It is significant that when the automobile first appeared, it was viewed as an answer to the increasingly serious urban-pollution problems created by the use of horses.

251. David Swinbanks, "Going for 'Green Technology,'" *Nature* 350 (Mar. 28, 1991): 266–67.

252. Lynton Keith Caldwell, *Between Two Worlds: Science, the Environmental Movement, and Policy Choice* (New York: Cambridge University Press, 1990).

253. Lynton Keith Caldwell, *International Environmental Policy: Emergence and Dimensions*, Duke Press Policy Studies (Durham, N.C.: Duke University Press, 1984).

254. "Protecting the Euro-Environment," *Geography* 74 (Jan. 1989): 47–52; Nigel Haigh, "The European Community and International Environmental Policy," *International Environmental Affairs* 3, no. 3 (Summer 1991): 163–80.

255. French, *Green Revolutions;* "East European Pollution: Clearing Up after Communism," *The Economist* Feb. 17, 1990, p. 54; Mark M. Nelson, "Darkness

at Noon: As Shroud of Secrecy Lifts in East Europe, Smog Shroud Emerges," *Wall Street Journal*, Mar. 1, 1990, p. 1.

256. Wade Rowland, *The Plot to Save the World: The Life and Times of the Stockholm Conference on the Human Environment* (Toronto: Clarke, Irwin, 1973).

257. Keith Bradsher, "U.S. Ban on Mexico Tuna Is Overruled," *New York Times*, Aug. 23, 1991, p. D1. For an excellent analysis of the interaction of international trade with environmental regulations, see Patrick Low and Raed Safadi, "Trade Policy and Pollution," paper prepared for Symposium on International Trade and the Environment, sponsored by the International Trade Division, International Economics Department, World Bank, Washington, D.C., Nov. 21–22, 1991.

258. Feshbach and Friendly, *Ecocide in the USSR*; French, *Green Revolutions*; Joseph Alcamo, Roderick Shaw, and Leen Hordijk, eds., *The RAINS Model of Acidification: Science and Strategies in Europe* (Hingham, Mass.: Kluwer, 1990).

259. Michael G. Renner, "Military Victory, Ecological Defeat," *World-Watch*, July/Aug. 1991, pp. 27–33.

260. The best general discussion of tightly coupled technological systems and their vulnerability to failure is to be found in Charles Perrow, *Normal Accidents: Living with High-Risk Technologies* (New York: Basic Books, 1984).

261. Philip Fites, *The Computer Virus Crisis* (New York: Van Nostrand Reinhold, 1989); James A. Schweitzer, *Managing Information Security: Administrative, Electronic, and Legal Measures to Protect Business Information*, 2d ed. (Boston: Butterworth, 1990).

262. G. Christopher Anderson, "Hacker Trial under Way," *Nature* 343 (Jan. 13, 1990): 200; John Markoff, "Arrests in Computer Break-in Show a Global Peril," *New York Times*, Apr. 4, 1990, p. 1; Mark Lewyn, "Hackers: Is a Cure Worse than the Disease? Blocking Their Mischief Could Mean Crippling Computer Networks," *Business Week*, Dec. 4, 1989, p. 371. *US News & World Report* alleged that U.S. intelligence used a computer virus to disable Iraqi air-defense computers during the 1991 Gulf War (*US News & World Report, Triumph without Victory: The Unreported History of the Gulf War* [New York: Times Books,]; "Computer Virus Use Cited in Gulf War," *Boston Globe*, Jan. 12, 1992, p. 12).

263. Computer scientist William Dowling reported in *Notices of the American Mathematical Society* (Sept. 1990) that there is no way, short of isolation, to protect a computer against all possible viral attacks: the existence of computer viruses is "an inevitable consequence of fundamental properties of any computing domain" (quoted in Barry Cipra, "Eternal Plague: Computer Viruses," *Science* 249 [Sept. 21, 1990]: 1381).

264. Eileen Shanahan, "F.P.C. Criticizes Power Systems in Nov. 9 Failure," *New York Times*, Dec. 7, 1965, p. 1.

265. Evelyn Richards, "A System on Overload; Our Unlimited Appetite for Software Strains Our Ability to Produce It," *Washington Post National Weekly Edition*, Dec. 31, 1990–Jan. 6, 1991, p. 6; Dennis Hevesi, "Blown Fuse Causes Big Phone Failure in New York City," *New York Times*, Sept. 18, 1991, p. 1.

266. William H. McNeill, "Control and Catastrophe in Human Affairs," in

A World to Make: Development in Perspective, Daedalus 118, no. 1 (Winter 1989): 12.

267. Office of Technology Assessment, *Global Standards: Building Blocks for the Future* (Washington, D.C.: OTA, Mar. 1992); John Burgess, "Making the World Safe for Square Pegs and Square Holes: European Product Standards are Critical for Americans," *Washington Post National Weekly Edition*, Dec. 9–15, 1991, p. 19.

268. Rhonda Joyce Crane, *The Politics of International Standards: France and the Color TV War* (Norwood, N.J.: Ablex, 1979).

269. Bob Francis, "Desktop Tug of War: The Operating System Choices the IS Executive Makes Today Will Shape the Character and Complexity of Corporate Applications for a Long Time to Come," *Datamation* 35, no. 4 (Feb. 15, 1989): 18–25; Gary McWilliams and Evan I. Schwartz, "Will This $150 Million Brainchild Be an Orphan?" *Business Week*, Oct. 29, 1990, p. 98; Andrew Pollack, "Microsoft Widens Its Split with IBM over Software," *New York Times*, July 27, 1991, p. 33.

270. Loretta Anania, "The Politics of Integration: Telecommunications Planning in the Information Societies" (Ph.D. diss., Massachusetts Institute of Technology, Department of Political Science, Feb. 1990).

271. "ISDN: France Leads the Way," *French Advances in Science and Technology* (Embassy of France in the United States) 2, no. 1 (Spring 1988): 4.

272. "U.S. Plans Next Electronics War," *Nature* 338 (Mar. 9, 1989): 106; John Burgess, "High-Definition TV Leaves Americans on Starting Block," *International Herald Tribune*, Mar. 8, 1989, p. 10; Office of Technology Assessment, *The Big Picture: HDTV and High Resolution Systems* (Washington, D.C.: OTA, June 12, 1990).

273. Alan G. Stoddard and Mark D. Dibner, "Europe's HDTV: Tuning Out Japan," *Technology Review* 92, no. 3 (Apr. 1989): 39.

274. Edmund L. Andrews, "U.S. Makes Gains in Race to Develop Advanced TV," *New York Times*, Dec. 21, 1990, p. 1.

275. David Hack, *High-Definition Television*, CRS Issue Brief, Congressional Research Service, Library of Congress (Washington, D.C., Oct. 6, 1989).

276. Hack, *High-Definition Television*, p. 6; Andrews, "U.S. Makes Gains."

277. Stoddard and Dibner, "Europe's HDTV."

278. Office of Technology Assessment, *The Big Picture*.

Chapter Five

1. John Tyndall recognized in 1861 that atmospheric variation could influence climate (Gordon J. MacDonald, "Scientific Basis for the Greenhouse Effect," *Journal of Policy Analysis and Management* 7, no. 3 [Spring 1988]: 427). The first major prediction of the global effects of carbon dioxide from fossil-fuel combustion was made by Svante Arrhenius ("On the Influence of Carbonic Acid in the Air upon the Temperature on the Ground," *Philosophical Magazine and Journal of Science* 41, no. 237 [1896]).

2. *The Limits to Growth*, in particular, excited enormous popular interest but dismayed responsible scientists because of its controversial methods and use of data (Dennis Meadows and Donella H. Meadows, *The Limits to Growth: A Report for the Club of Rome's Project on the Predicament of Mankind* [New York: Uni-

verse Books, 1972]). Effective critiques were offered in Wilfred Beckerman, *In Defence of Economic Growth* (London: Jonathan Cape, 1974); and Carl Kaysen, "The Computer That Cried W*O*L*F," *Foreign Affairs* 50, no. 4 (July 1972): 660–68. See also H.S.D. Cole et al., eds., *Models of Doom: A Critique of the Limits to Growth* (New York: Universe Books, 1973). The report of the Study of Critical Environmental Problems (SCEP), organized by Carroll Wilson of MIT, was an important contribution to a responsible recognition of the issue (*Man's Impact on the Global Environment: Assessment and Recommendations for Action* [Cambridge, Mass.: MIT Press, 1970]).

3. For example, the U.S. proposed global-change research budget for the 1993 fiscal year was $1.37 billion, more than twice the budget of two years earlier (Office of Science and Technology Policy, Committee on Earth and Planetary Sciences, Federal Coordinating Council for Science, Engineering and Technology, *Our Changing Planet: The FY 1993 U.S. Global Change Research Program* [Washington, D.C., 1992], p. 28).

4. Among the first scholars of international relations to deal with the issue were Harold H. and Margaret T. Sprout (*Toward a Politics of the Planet Earth* [New York: Van Nostrand Reinhold, 1971]). One of the most interesting early analyses of the international political aspects of global change was by Crispin Tickell (*Climate Change and World Affairs* [Cambridge, Mass.: Harvard Center for International Affairs, 1977], later reissued in a revised edition [Lanham, Md.: University Press of America, 1986]). Tickell was a British Foreign Service officer and became a key adviser to Prime Minister Thatcher on this issue. See also Eugene B. Skolnikoff, "The Policy Gridlock on Global Warming," *Foreign Policy*, no. 79 (Summer 1990): 77–93; and "Symposium on Global Climate Change and Public Policy," *Policy Studies Journal* 19, no. 2 (Spring 1991): 43–161.

5. The basic facts of the greenhouse effect are covered in an increasing number of articles and books. The panels of the Committee on Science, Engineering, and Public Policy of the U.S. National Academy of Sciences, National Academy of Engineering, and Institute of Medicine that reported in 1991 and 1992 included many of the natural and social scientists and engineers most knowledgeable on the subject. Their reports offer the most authoritative summary of the state of knowledge to that time and will be drawn on heavily in this chapter. There were four panels: on effects, mitigation, adaptation, and synthesis. The results were published in two volumes, *Policy Implications of Greenhouse Warming: The Synthesis Panel* (Washington, D.C.: National Academy Press, 1991) and *Policy Implications of Greenhouse Warming: Mitigation, Adaptation, and the Science Base* (Washington, D.C.: National Academy Press, 1992). *The Synthesis Panel* provides the general conclusions; the other volume encompasses the full study. In addition, two of the panels released prepublication reports in 1991; they will be referred to here as NAS, *Mitigation Panel* and NAS, *Adaptation Panel*. See also MacDonald, "Scientific Basis," pp. 425–44; and Norman J. Rosenberg et al., eds., *Greenhouse Warming: Abatement and Adaptation* (Washington, D.C.: Resources for the Future, 1989).

6. Water vapor is a major factor in the greenhouse effect; its effects entail many uncertainties and important but still-unclear feedback relationships, particularly with respect to the effects of clouds. Its role in atmospheric warming is affected by so many other factors in addition to human activities, however, that it is not usually included in analyses of greenhouse gases.

7. NAS, *Synthesis Panel*, p. 3.

8. Ibid., p. 10.

9. This uncertainty in the globe's carbon budget complicates the calculation of how much reduction of CO_2 emissions would be necessary to "stabilize" the temperature. Even capping emissions at the 1990 rate would not necessarily stabilize the atmospheric concentration of CO_2, depending on what proportion of those emissions was absorbed, and at what rate, in the oceans and other carbon sinks.

10. Stephen H. Schneider, "The Changing Climate," *Scientific American* 261, no. 3 (Sept. 1989): 73.

11. WMO/UNEP Intergovernmental Panel on Climate Change (IPCC), *1992 IPCC Supplement* (Geneva, Feb. 1992), p. 21. David Victor shows how uncertain the calculations of "equivalence" among gases can be and how many variables are involved in "Calculating Greenhouse Budgets," *Nature* 347 (Oct. 4, 1990): 431.

12. An excellent history of the Montreal Protocol negotiations is provided in Richard Elliot Benedick, *Ozone Diplomacy: New Directions in Safeguarding the Planet* (Cambridge, Mass.: Harvard University Press, 1991).

13. William J. Cromie, "Increase in Ozone-Destroyers Found: MIT Network Measures Buildup of Gases," *MIT Report*, May 1989, pp. 4–5; Keith Shine, "Effects of CFC Substitutes," *Nature* 344 (Apr. 5, 1990): 492–93.

14. Michael B. McElroy, "The Challenge of Global Change," *Bulletin of the American Academy of Arts and Sciences* 42, no. 5 (Feb. 1989): 28.

15. MacDonald, "Scientific Basis," p. 434.

16. H. Rohde, "A Comparison of the Contribution of Various Gases to the Greenhouse Effect," *Science* 248 (June 8, 1990): 1217–19; David G. Victor, "Leaking Methane from Natural Gas Vehicles: Implications for Transportation Policy in the Greenhouse Era," *Climatic Change* 20 (Feb. 1992): 113–41.

17. MacDonald, "Scientific Basis," p. 435.

18. Recognizing that the comparisons among gases are quite complex and subject to many uncertainties, the NAS report used representative values for simplicity.

19. NAS, *Synthesis Panel*, p. 25; *Earthquest* (U.S. Department of Energy, Office of Interdisciplinary Studies) 5, no. 1 (Spring 1991). The average global temperature during the Ice Age extreme eighteen thousand years ago, when ice to a depth of several kilometers covered North America and Europe, was only 5° colder than today (MacDonald, "Scientific Basis," p. 440).

20. NAS, *Synthesis Panel*, p. 25.

21. WMO/UNEP Intergovernmental Panel on Climate Change (IPCC), *Scientific Assessment of Climate Change*, Report of Working Group I, Policymakers' Summary (Geneva, July 1990), p. 2.

22. NAS, *Synthesis Panel*, pp. 36–40; Stephen H. Schneider, *Global Warming: Are We Entering the Greenhouse Century?* (San Francisco: Sierra Club Books, 1989), pp. 132–90; *Managing Planet Earth*, special issue, *Scientific American* 261, no. 3 (Sept. 1989).

23. NAS, *Synthesis Panel*, p. 26; IPCC, *Scientific Assessment*, p. 22.

24. If the models are correct, warming in North America would occur at a rate "ten to fifty times faster than the average climate change from the waning of the last ice age to the present!" (Rosenberg et al., *Greenhouse Warming*, p. 31). See also Wallace S. Broecker, "Unpleasant Surprises in the Greenhouse," *Nature* 328 (July

9, 1987): 123–26; and Terence Hughes, "Is the West Antarctic Ice Sheet Disintegrating?" *Journal of Geographical Research* 78 (1973): 7889–910.

25. Some representative examples (in addition to those already cited) of a rapidly growing body of books, reports, and articles are Jessica Tuchman Mathews, ed., *Preserving the Global Environment: The Challenge of Shared Leadership* (New York: W. W. Norton, 1991): F. Herbert Bormann and Stephen R. Kellert, eds., *Ecology, Economics, Ethics: The Broken Circle* (New Haven, Conn.: Yale University Press, 1991); and Albert Gore, *Earth in Balance: Ecology and the Human Spirit* (Boston: Houghton Mifflin, 1992).

26. The IPCC Scientific Assessment Panel indicated that up to a 60 percent reduction in emissions of CO_2 would be required to *stabilize* concentrations at 1990 levels (IPCC, *Scientific Assessment*, pp. xi, xviii).

27. NAS, *Synthesis Panel*, p. 63; William D. Nordhaus, "Global Warming: Slowing the Greenhouse Express," in Henry J. Aaron, ed., *Setting National Priorities: Policy for the Nineties* (Washington, D.C.: Brookings Institution, 1990), pp. 185–211; idem, "The Cost of Slowing Climate Change: A Survey," *Energy Journal* 12 (1991): 37–65.

28. Rohde, "Comparison"; IPCC, *Scientific Assessment*, p. 7.

29. For a discussion of the problems of dealing with long time horizons, see Jerome Rothenberg, "Intertemporal and Inter-Generational Anomalies for International Response to Global Climate Change," in Nazli Choucri, ed., *Global Change: Environmental Challenges and International Responses* (Cambridge, Mass.: MIT Press, forthcoming).

30. Broecker, "Unpleasant Surprises in the Greenhouse." It is also possible that climate change may be chaotic; that is, there could be substantially different climatic responses as a result of small differences in initial conditions. If so, the actual details of changes in climate due to the greenhouse effect could be for all practical purposes unpredictable (James Gleick, "Making Sense of Nature's Mess," *Washington Post*, Oct. 18, 1987, p. D3; Alan McRobie and Michael Thompson, "Chaos, Catastrophes and Engineering," *New Scientist* 126 [June 9, 1990]: 41–46).

31. Benedick, *Ozone Diplomacy*, p. 108; Paul Brodeur, "In the Face of Doubt," *New Yorker*, June 9, 1986, pp. 83–84.

32. "Fossil Fuels CO_2 Emissions: Three Countries Account for 50 percent in 1986," *CDIAC Communications* (Carbon Dioxide Information Analysis Center, Oak Ridge National Laboratory), Winter 1989, p. 1.

33. Arrhenius, "Carbonic Acid."

34. MacDonald, "Scientific Basis," p. 430.

35. Schneider, *Global Warming*, p. 40.

36. The IPCC assessment concluded that the global mean surface air temperature had increased 0.3 to 0.6°C over the last one hundred years (IPCC, *Scientific Assessment*, p. 3). Other data is ambiguous; see "Has the Globe Really Warmed?" MIT Reporter, *Technology Review* 92, no. 8 (Nov./Dec. 1989): 80; A. S. McLaren, R. G. Barry, and R. H. Bourke, "Could Arctic Ice Be Thinning?" *Nature* 345 (June 28, 1990): 762; Richard A. Kerr, "Could the Sun Be Warming the Climate?" *Science* 254 (Nov. 1, 1991): 254–55; Leslie Roberts, "Warm Waters, Bleached Corals," *Science* 250 (Oct. 12, 1990): 213; William K. Stevens, "Warming in North Disrupts Ecology," *New York Times*, Nov. 16, 1990, p. A30; Peter

Aldhous, "1990 Warmest Year on Record," *Nature* 349 (Jan. 17, 1991): 186; and John E. Walsh, "The Arctic as a Bellwether," *Nature* 352 (July 4, 1991): 19–20.

37. David A. Worth and Daniel A. Lashof, "Beyond Vienna and Montreal: Multilateral Agreements on Greenhouse Gases," *Ambio* 19, no. 6–7 (Oct. 1990): 305–10; Paul Lewis, "Thatcher Urges Pact on Climate," *New York Times*, Nov. 9, 1989, p. A17; Allen L. Hammond, Eric Rodenburg, and William R. Moomaw, "Calculating National Accountability for Climate Change," *Environment* 33, no. 1 (Jan./Feb. 1991): 15.

38. There is a higher probability that so-called no-regrets, or insurance, measures will be taken—those that contribute to reducing greenhouse-gas emissions but that also can serve other objectives, such as energy conservation, improved energy efficiency, or reduced pollution. Their role in reducing the presumed danger of global warming may be a useful selling point, but in the absence of a clear case for committing resources for that goal, the no-regrets measures must be justified on their own merits.

39. NAS, *Synthesis Panel*, p. 45. One member of the panel, Jessica Mathews, strenuously disagreed with that conclusion, arguing it was unwarranted given the extent of the uncertainties and the rapidity and size of a possible temperature increase.

40. Hammond, Rodenburg, and Moomaw, "National Accountability," p. 15; Karen Schmidt, "How Industrial Countries Are Responding to Global Climate Change," *International Environmental Affairs* 3, no. 4 (Fall 1991): 292–317.

41. William K. Stevens, "White House Vows Action to Cut U.S. Global Warming Gases," *New York Times*, Feb. 28, 1992, p. A6.

42. Benedick, *Ozone Diplomacy*, pp. 118–28; Keith Schneider, "Bush Orders End to Ozone Destroyers by 1996," *New York Times*, Feb. 12, 1992, p. A18.

43. Dr. James Hansen, director of NASA's Goddard Institute for Space Studies, began testimony before Congress on June 23, 1988, with the statement "Global warming has begun" (quoted in Christopher Flavin, "The Heat Is On," *World-Watch*, Nov./Dec. 1988, p. 10).

44. NAS, *Mitigation Panel*, chap. 10.

45. Broecker, "Unpleasant Surprises in the Greenhouse," pp. 124–25; NAS, *Synthesis Panel*, p. 24.

46. Scientists are divided over the level of threat. A 1989 study of over four hundred American scientists showed that 58 percent supported "decisive action to respond to the greenhouse effect" (John Doble, Amy Richardson, and Allen Danks, *Global Warming Caused by the Greenhouse Effect*, vol. 3 of *Science and the Public* [New York: Public Agenda Foundation, 1990]). The 1990 IPCC studies do reflect a consensus among leading scientists about the state of scientific knowledge, as do the 1991 National Academy of Sciences panel reports; important differences remain, however, about uncertainties, reliability of the models, and implications for early action (IPCC, *Scientific Assessment;* NAS, *Synthesis Panel;* NAS, *Mitigation Panel*).

47. Lynton Keith Caldwell, *Between Two Worlds: Science, the Environmental Movement, and Policy Choice* (New York: Cambridge University Press, 1990), pp. 105–23.

48. Edith Brown Weiss, *In Fairness to Future Generations: International Law, Common Patrimony, and Intergenerational Equity* (Dobbs Ferry, N.Y.: Transnational Publishers, 1989).

49. Bill McKibben, *The End of Nature* (New York: Random House, 1989).

50. NAS, *Mitigation Panel*, chap. 8.

51. Richard H. Ullman, "Redefining Security," *International Security* 8, no. 1 (Summer 1983): 133.

52. "A Plague on Us," *The Economist*, Jan. 24, 1987, pp. 41–42; Lori Heise, "AIDS: New Threat to the Third World," *World-Watch*, Jan./Feb. 1988, pp. 19–27; Barbara J. Culliton, "AIDS against the Rest of the World," *Nature* 352 (July 4, 1991): 15; "AIDS: Poor Man's Plague," *The Economist*, Sept. 21, 1991, pp. 21–24.

53. William H. McNeill, *Plagues and Peoples* (Garden City, N.Y.: Anchor Press, 1976); Barbara W. Tuchman, *A Distant Mirror: The Calamitous Fourteenth Century* (New York: Alfred A. Knopf, 1978).

54. Philip S. Gutis, "Troubled Seas: Global Red Tides of Algae Bring New Fears," *New York Times*, May 3, 1988, p. C1.

55. Michael Richardson, "Asia and Pacific Face a Seafood Crisis," *International Herald Tribune*, Apr. 10, 1989, p. 4; Sandra Postel, "Land's End," *World-Watch*, May/June 1989, p. 12; NAS, *Mitigation Panel*, chap. 9; Alan Randall, "The Value of Biodiversity," *Ambio* 20, no. 2 (Apr. 1991): 64–68.

56. Warren E. Leary, "Earth Had a Close Call with a Large Asteroid," *International Herald Tribune*, Apr. 21, 1989; William J. Broad, "There's a Doomsday Rock; But When Will It Strike?" *New York Times*, June 18, 1991, p. C1; American Institute of Aeronautics and Astronautics, *Dealing with the Threat of an Asteroid Striking the Earth*, an AIAA position paper (Washington, D.C., Apr. 1990).

57. Duncan Steel, "Our Asteroid-Pelted Planet," *Nature* 354 (Nov. 28, 1991): 265–67.

58. Robert Matthews, "A Rocky Watch for Earthbound Asteroids," *Science* 255 (Mar. 1992): 1204.

59. The U.S. National Aeronautics and Space Administration (NASA) is considering what would be required to establish a monitoring network (Matthews, "A Rocky Watch").

Chapter Six

1. The focus in this chapter is on the detailed problems of governance that are altering the functions and processes of governments and international organizations. The larger, overarching changes with respect to, for example, the autonomy of the state are considered in the concluding chapter.

2. In postmortems after the Gulf War, Pentagon officials expressed "amazement . . . about how quickly warfare is being overtaken by technologies that compress time and therefore affect human judgments" (Patrick E. Tyler, "Pentagon Reassesses the Vulnerability of Its Tanks," *New York Times*, Aug. 15, 1991, p. A10).

3. Woodrow Wilson Center for International Scholars, *An Evening with Harold Macmillan at the Wilson Center* (Washington, D.C., 1986), p. 8.

4. Theodore C. Sorenson, "The President and the Secretary of State," *Foreign Affairs* 66, no. 2 (Winter 1987/88): 236.

5. For a lucid discussion of the problems of comparing the future with the present when considering investments to deal with global warming, see National Academy of Sciences, *Policy Implications of Global Warming—The Synthesis Panel* (Washington, D.C.: National Academy Press, 1991), pp. 11–12. See also Jerome Rothenberg, "Intertemporal and Inter-Generational Anomalies for International Response to Global Climate Change," in Nazli Choucri, ed., *Global Change: Environmental Challenges and International Responses* (Cambridge, Mass.: MIT Press, forthcoming).

6. Robert C. Lind et al., eds., *Discounting for Time and Risk in Energy Policy* (Baltimore: Johns Hopkins University Press, 1982).

7. There is a substantial literature on the interaction of science and technology with foreign policy and on the roles of scientific experts in policy formation. See particularly Eugene B. Skolnikoff, *Science, Technology, and American Foreign Policy* (Cambridge, Mass.: MIT Press, 1967); Sanford Lakoff, *Knowledge and Power: Essays on Science and Government* (New York: Free Press, 1966); Joseph S. Szyliowicz, ed., *Technology and International Affairs* (New York: Praeger, 1981); and Senate Commission on the Organization of the Government for the Conduct of Foreign Policy, *Appendices*, vol. 1, *Appendix B: The Management of Global Issues* (Washington, D.C., June 1975).

8. The general problem of inadequate understanding of science and technology in foreign offices, and the implications of that deficiency, are discussed in Skolnikoff, *Science, Technology, and American Foreign Policy*, pp. 249–98. A report of the Carnegie Commission on Science, Technology, and Government analyzes in detail the situation and needs in the U.S. government (*Science and Technology in U.S. International Affairs: Challenges for Twenty-first Century Governmental Policy and Organization* [New York: Carnegie Commission on Science, Technology, and Government, Jan. 1992]).

9. It is worth noting that it is not only those without technical training who are vulnerable to overvaluing technology's potential contribution. Scientists and engineers have proven to be culpable as well. The most egregious example is the overselling of SDI by a number of scientists (see chapter 3). The attention accorded the well-orchestrated promise of unlimited energy from cold fusion provides another illustration (Frank Close, *Too Hot to Handle: The Race for Cold Fusion* [Princeton: Princeton University Press, 1991]).

10. The latest study is by the prestigious Carnegie Commission on Science, Technology, and Government, which once again lays out the elements of the issue and makes recommendations for change (Carnegie Commission, *Science and Technology in U.S. International Affairs*). The endorsement of the report at the time of its release by distinguished senior officials of previous administrations (e.g., Secretary of State George Shultz, Deputy Secretary of State John C. Whitehead, Director of the Foreign Service Harry Barnes, and National Security Adviser McGeorge Bundy) and the now-evident significance of science and technology in foreign policy may lead to more attention to the report's recommendations. Past experience offers little encouragement, however.

11. Skolnikoff, *Science, Technology, and American Foreign Policy*, p. 267.

12. Eugene B. Skolnikoff, "Computers, Armaments, and Stability," in Denis P. Donnelly, ed., *The Computer Culture: A Symposium to Explore the Computer's Impact on Society* (Rutherford, N.J.: Fairleigh Dickinson University Press, 1985), pp. 124–35.

13. Thomas F. Malone and Thomas Rosswall, "Global Change and the IGBP," *Science International Newsletter: Special Issue*, Sept. 1991, pp. 50–51.

14. Robert C. Wood, "Scientists and Politics: The Rise of an Apolitical Elite," in Robert Gilpin and Christopher Wright, eds., *Scientists and National Policy-Making* (New York: Columbia University Press, 1964), pp. 41–72.

15. Skolnikoff, *Science, Technology, and American Foreign Policy*, pp. 249–98.

16. M. Stephen Weatherford, "The International Economy as a Constraint on U.S. Macroeconomic Policymaking," *International Organization* 42, no. 4 (Autumn 1988): 605.

17. The leak of a draft U.S. Department of Defense policy statement gives substance to the fact that a resurgence of hegemonic power, rather than collective action, to deal with international issues indeed has some reality. The draft—which does not have the status of an official U.S. policy position—postulated a continuing dominant role for the United States as the single remaining superpower, whose position would be perpetuated by constructive behavior and which would actively prevent the emergence of competing power centers. Apparently, there is little or no mention in the document of collective action through the UN (Patrick E. Tyler, "U.S. Strategy Plan Calls for Insuring No Rivals Develop," *New York Times*, Mar. 8, 1992, p. 1).

18. Mitchel B. Wallerstein, ed., *Scientific and Technological Cooperation among Industrialized Countries* (Washington, D.C.: National Academy Press, 1984); Alexander Keynan, *The United States as a Partner in Scientific and Technological Cooperation: Some Perspectives from across the Atlantic*, (New York: Carnegie Commission on Science, Technology, and Government, June 1991).

19. U.S. House of Representatives, Committee on Foreign Affairs and Committee on Science and Technology, *Annual Reports Submitted to the Congress by the President, Pursuant to Section 503(b) of Title V of PL 95–426*, Joint Committee Prints.

20. Eugene B. Skolnikoff, "Problems in the U.S. Government Organization and Policy Process for International Cooperation in Science and Technology," in Wallerstein, *Scientific and Technological Cooperation*, pp. 29–43.

21. Paul M. Kennedy, *The Rise and Fall of the Great Powers: Economic Change and Military Conflict from 1500 to 2000* (New York: Random House, 1987); Joseph S. Nye, Jr., *Bound to Lead: The Changing Nature of American Power* (New York: Basic Books, 1990). See also chapter 4.

22. Keynan, *United States as a Partner*, pp. 41–60.

23. Harold K. Jacobson, *Networks of Interdependence: International Organizations and the Global Political System*, 2d ed. (New York: Alfred A. Knopf, 1984), pp. 30–58; Mark W. Zacher, "The Decaying Pillars of the Westphalian Temple: Implications for International Order and Governance," in James N. Rosenau and Ernst-Otto Czempiel, eds., *Governance without Government: Order and Change in World Poltics* (New York: Cambridge University Press, 1992), pp. 58–101.

24. Jacobson, *Networks of Interdependence*, pp. 47–52; see also Clive Archer,

International Organizations (London: Allen & Unwin, 1983), p. 130; and Zacher, "Decaying Pillars," p. 6. Different authors have different absolute numbers of IGOs and NGOs, depending on definitions; the proportions are similar, however.

25. There is an extensive literature on international organizations—their origins, functions, and political relationships. See particularly Jacobson, *Networks of Interdependence;* David Mitrany, *A Working Peace System* (Chicago: Quadrangle Books, 1966); Ernst B. Haas, *Beyond the Nation-State: Functionalism and International Organizations* (Palo Alto, Calif.: Stanford University Press, 1964); idem, *When Knowledge Is Power: Three Models of Change in International Organizations* (Berkeley: University of California Press, 1990); and Archer, *International Organizations.* For an analysis focused on the role of science and technology in those organizations, see Eugene B. Skolnikoff, *The International Imperatives of Technology: Technological Development and the International Political System,* Research Series, no. 16, Institute of International Studies, University of California, Berkeley, 1972.

26. Substantial theoretical schools were based on this general idea, growing largely from the work of David Mitrany (*A Working Peace System*) and, later, Ernst Haas (*Beyond the Nation-State*) and Joseph S. Nye (*Peace in Parts: Integration and Conflict in Regional Organization* [Boston: Little, Brown, 1971]).

27. Skolnikoff, *International Imperatives,* pp. 132–48.

28. For an extended analysis of how international organizations change over time, see Haas, *When Knowledge Is Power.*

29. Skolnikoff, *International Imperatives,* pp. 158–66.

30. McGeorge Bundy, *Danger and Survival: Choices about the Bomb in the First Fifty Years* (New York: Random House, 1988), chap. 4.

31. Trevor Rowe, "Tough UN Plan Proposed to Monitor Iraqi Weapons," *Boston Globe,* Aug. 3, 1991, p. 2; Paul Lewis, "Iraq Agrees to Destroy Missiles to Meet U.N. Cease-Fire Terms," *New York Times,* Mar. 21, 1992, p. 1; Overseas Development Council, *Humanitarian Intervention in a New World Order,* ODC Policy Focus No. 1 (Washington, D.C.: ODC, 1992).

32. Gregory F. Treverton, ed., *The Shape of the New Europe* (New York: Council on Foreign Relations, 1992); Neill Nugent, *The Government and Politics of the European Community* (Durham, N.C.: Duke University Press, 1989); Jeffrey Harrop, *The Political Economy of Integration in the European Community* (Brookfield, Vt.: Gower Publishing, 1989).

33. "A Light Is Dimmed," *The Economist,* Jan. 19, 1991, p. 48.

34. Zacher concludes that the number of international NGOs in fact grew much faster than the number of IGOs in the postwar period ("Decaying Pillars," p. 6). Archer estimates there will be close to ten thousand international NGOs by the end of the twentieth century (*International Organizations,* p. 130).

35. J. J. Lador-Lederer, *International Non-Governmental Organizations and Economic Entities: A Study in Autonomous Organization and* Ius Gentium (Leyden, the Netherlands: A. W. Sythoff, 1963).

36. National Research Council, *The Revolution in Information and Communications Technology and the Conduct of U.S. Foreign Affairs,* summary report of the workshop held at the National Academy of Sciences, Washington, D.C., Sept. 14–15, 1987 (Washington, D.C.: National Academy Press, 1988), pp. 31–35. The

Earth Observing System Data and Information System NASA is establishing to process data from new earth-observation satellites to be established to monitor global change will, by the year 2000, contain one thousand times the amount of text stored in the Library of Congress and will take in two billion bits of data every day (Eliot Marshall, "Accountants Fret over EOS Data," *Science* 255 [Mar. 6, 1992]: 1206).

37. Climate data from U.S. satellites and weather stations is accumulating at a rate of more than one thousand magnetic tapes a day, a flow that will increase archives a hundredfold by 2000 (Christopher Anderson, "Too Much of a Good Thing," *Nature* 349 [Feb. 14, 1991]: 553). The U.S. Global Change Research Program will greatly increase this already-massive data flow (Office of Science and Technology Policy, Committee on Earth and Planetary Sciences, Federal Coordinating Council for Science, Engineering and Technology [FCCSET], *Our Changing Planet: The FY 1993 U.S. Global Change Research Program* [Washington, D.C., 1992], pp. 61–64).

38. Michael Grubb, *The Greenhouse Effect: Negotiating Targets* (London: Royal Institute of International Affairs, 1989) pp. 27–42.

39. James N. Rosenau discusses this point in "The State in an Era of Cascading Politics," *Comparative Political Studies* 21, no. 1 (Apr., 1988): 13–44.

Chapter Seven

1. Stephen D. Krasner, *International Regimes* (Ithaca, N.Y.: Cornell University Press, 1983), p. 366.

2. Janice E. Thomson and Stephen D. Krasner, "Global Transactions and the Consolidation of Sovereignty," in Ernst-Otto Czempiel and James N. Rosenau, eds., *Global Changes and Theoretical Challenges: Approaches to World Politics for the 1990s* (Lexington, Mass.: Lexington Books, 1989), p. 196. For the most cogent statement of the realist position, see Kenneth N. Waltz, *Theory of International Politics* (Reading, Mass.: Addison-Wesley, 1979).

3. Waltz, *Theory of International Politics*, pp. 104–7, 210.

4. James N. Rosenau, *Turbulence in World Politics: A Theory of Change and Continuity* (Princeton: Princeton University Press, 1990), p. 71.

5. Stanley Hoffmann, "War and the Future of International Relations," in Jean-Jacques Salomon, ed., *Science, War, and Peace* (Paris: Economica, 1990), pp. 177–90.

6. The problem of accommodating and measuring change is one of the more difficult issues in various theoretical schools. See particularly John Gerard Ruggie, "Continuity and Transformation in the World Polity: Toward a Neorealist Synthesis," *World Politics* 35, no. 2 (Jan. 1983): 261–85; Robert O. Keohane, ed., *Neorealism and Its Critics* (New York: Columbia University Press, 1986); Rosenau, *Turbulence in World Politics*; Mark W. Zacher, "The Decaying Pillars of the Westphalian Temple: Implications for International Order and Governance," in James N. Rosenau and Ernst-Otto Czempiel, eds., *Governance without Government: Order and Change in World Politics* (New York: Cambridge University Press, 1992), pp. 58–101; and Robert Gilpin, *War and Change in World Politics* (New York: Cambridge University Press, 1981).

7. Krasner, *International Regimes*, p. 7.

8. Regimes can be defined as "networks of rules, norms, and procedures that regularize behavior and control its effects" (Robert O. Keohane and Joseph S. Nye, Jr., *Power and Interdependence: World Politics in Transition* [Boston: Little, Brown, 1977], p. 19). See also Robert O. Keohane, "Theory of World Politics: Structural Realism and Beyond," in Keohane, *Neorealism and Its Critics* (pp. 158–203; and Krasner, *International Regimes*, p. 2.

9. Rosenau, *Turbulence in World Politics*, pp. 96–97.

10. Vaclav Havel, playwright and past president of Czechoslovakia, goes further, arguing that the modern age, characterized by a belief—grounded in science—in a knowable world governed by universal laws, is no longer able to cope with the problems civilization faces: "We treat the fatal consequences of technology as though they were a technical defect that could be remedied by technology alone. We are looking for an objective way out of a crisis of objectivism." He goes on to argue for a "postmodern" politics that will require fundamental changes in behavior and that will be necessary to bring about systemic and institutional changes. But he does not spell out what those changes, either in behavior or institutions, may be (Vaclav Havel, "The End of the Modern Era," *New York Times*, Mar. 1, 1992, p. 15).

11. An excellent collection of essays about scientific and technological futures that includes important selected references is Joseph J. Corn, ed., *Imagining Tomorrow: History, Technology, and the American Future* (Cambridge, Mass.: MIT Press, 1986).

12. Philip Ball and Laura Garwin, "Science at the Atomic Scale," *Nature* 355 (Feb. 27, 1992): 761–66; Office of Technology Assessment, *Miniaturization Technologies* (Washington, D.C.: OTA, Nov. 1991).

13. Quoted by D. Allan Bromley, Assistant to the President for Science and Technology, at the National Press Club, Oct. 23, 1990.

14. Some developments may mimic life in ways that gradually (and disturbingly) blur the distinctiveness of what we now consider biological life—for example, computer programs with characteristics of reproduction, mutation, and genetic change that satisfy standard definitions of life (Malcolm W. Browne, "Lively Computer Creation Blurs Definition of Life," *New York Times*, Aug. 27, 1991, p. C1).

15. An excellent summary of the situation in the United States can be found in Gerald Holton, "How to Think about the 'Anti-Science' Phenomenon," paper prepared for a joint U.S.-USSR workshop, "Anti-Science and Anti-Technology Trends in the U.S. and U.S.S.R.," Massachusetts Institute of Technology, Cambridge, Mass., May 2–3, 1991.

16. "The Edge of Ignorance: A Survey of Science," *The Economist*, Feb. 16, 1991, p. 22.

Index